Automation, Collaboration, & E-Services

Volume 15

AF902993

The Automation, Collaboration, & E-Services series (ACES) publishes new developments and advances in the fields of Automation, collaboration and e-services; rapidly and informally but with a high quality. It captures the scientific and engineering theories and techniques addressing challenges of the megatrends of automation, and collaboration. These trends, defining the scope of the ACES Series, are evident with wireless communication, Internetworking, multi-agent systems, sensor networks, cyber-physical collaborative systems, interactive-collaborative devices, and social robotics – all enabled by collaborative e-Services. Within the scope of the series are monographs, edited volumes, lecture notes, and selected contributions from specialized conferences and workshops.

Puwadol Oak Dusadeerungsikul · Shimon Y. Nof

Cyber-Collaborative Algorithms and Protocols

Optimizing Agricultural Robotics

Springer

Puwadol Oak Dusadeerungsikul
Department of Industrial Engineering
Faculty of Engineering
Chulalongkorn University
Bangkok, Thailand

Shimon Y. Nof
PRISM Center and School of IE
Purdue University
West Lafayette, Indiana, USA

ISSN 2193-472X ISSN 2193-4738 (electronic)
Automation, Collaboration, & E-Services
ISBN 978-3-031-56195-5 ISBN 978-3-031-56196-2 (eBook)
https://doi.org/10.1007/978-3-031-56196-2

This Springer imprint is published by the registered company Springer Nature Switzerland AG
The registered company address is: Gewerbestrasse 11, 6330 Cham, Switzerland

Paper in this product is recyclable.

To my wife, Crystal

—Puwadol Oak Dusadeerungsikul

To my wife, Nava

—Shimon Y. Nof

…and to all our Purdue PRISM Center and PGRN, PRISM Global Research Network scholars who contributed to this knowledge.

Foreword

When I was invited to write a foreword for this book, I was deeply excited. With experience of several decades in agricultural engineering and agricultural robotics, I am grateful for the opportunity to express my insights on the visionary and yet necessary topic of this book: *Cyber-Collaborative Algorithms and Protocols: Optimizing Agricultural Robotics*, which this book addresses.

Throughout the course of human history, agriculture has maintained profound significance and served as the cornerstone of civilization. Its importance extends beyond the feeding needs of a growing population. It also encompasses the production of medicines, fiber, and fuel. Through the continual progression of scientific knowledge, technological innovation, and equipment refinement, the introduction of agrochemicals and genetically modified food into agriculture aims to attain maximal yields of reliable, high quality products, while reducing labor and energy costs. These innovations align with the goal of maximizing productivity and efficiency in agricultural practices.

During recent years, agriculture is undergoing the so-called fourth revolution. It is driven by significant advances in four technological domains: Information and Communications Technologies (ICT); Precision Agriculture (PA); computation (including Cyber, AI, and big data); and agricultural robotics. The overarching objective is to enhance both the quality and quantity of crops while minimizing the utilization of natural resources, attaining resilience, and optimizing the usage of human labor. Task Administration Protocols (TAPs) and Cyber-Collaborative Protocols (CCPs) of a Cyber-Physical System (CPS) are, in a way, the "missing link": They are designed to harness, integrate, and make the new four technological domains work effectively in agriculture.

This inspiring book offers an extensive background, overview, fundamental principles, and practical implementations of Cyber-Collaborative Protocols (CCPs) within Cyber-Physical Systems (CPS). These tools are important in many fields of work and service. The implementation in agricultural automation and robotics adds meaningful clarity. May this knowledge serve as a guiding beacon in this exciting journey.

Avital Bechar
The Institute of Agricultural
Engineering, Volcani Institute
Rishon LeTsiyon, Israel

Preface

We are inspired to write this book by a simple observation: in today's world, systems are becoming increasingly interconnected and, at the same time, more complex. These evolving characteristics of systems create key challenges to effective management and control. Moreover, technological advancement has led to agents in modern systems becoming more distributed and decentralized, further complicating the task of maintaining rational control, guidance, and oversight.

These challenges in modern systems call for the exploration of new management strategies. The need becomes especially evident in sectors like agriculture, where precision, efficiency, responsiveness, and adaptability are not just desirable but also required for long-term sustainability and, eventually, solving a bigger problem, food security problems, which have become more severe nowadays.

The most challenging context as an agricultural system is where the concept of Cyber-Collaborative Protocol (CCP), which offers a new approach to managing complex tasks and systems with collaboration, efficiency, and intelligence, truly shines. This book aims to translate the intricacies and potential of CCP into practice, with a particular focus on modern complex systems such as Agricultural Robotics Systems (ARS). We focus on providing a balanced mix of theoretical foundations, practical implementations, and forward-looking perspectives on the future of collaborative systems in the cyber age. Our exploration of CCP goes beyond technical intricacies; we also delve into real-world implications. Through our primary case study, ARS, we highlight the power, potential, and capability of CCP in various levels of collaborative systems.

Importantly, we would like to express our gratitude to our colleagues, contributors, and collaborators whose invaluable insights have made this book a reality. We acknowledge the support received from the PRISM Center at Purdue University; BARD Grant IS-4886-16R for "Development of a robotic inspection system for early identification and locating of biotic and abiotic stresses in greenhouse crops," PI A. Bechar, and by NSF Grant #1839971: "Pre-Skilling Workers, Understanding Labor Force Implications and Designing Future Factory Human-Robot Workflows Using Physical Simulation Platform," PI K. Ramani. We also thank our friendly Springer staff members who have helped us prepare and bring this book to you. We trust that our readers will discover this book to be both instructive and inspiring in their respective fields.

Lastly, welcome all readers to the world of Cyber-Collaborative Protocols, where humans and systems collaboration meets science and technology.

Bangkok, Thailand Puwadol Oak Dusadeerungsikul
West Lafayette, Indiana, USA Shimon Y. Nof

Contents

Chapter 1
Introduction: Cyber-Collaborative Protocols

Abstract Just as a good concert cannot be performed without a conductor, and a computer cannot work without an operating system, every collaborative system must be guided and administered to accomplish its goals. Task Administration Protocols (TAPs) have been developed to administer and manage complex, automated workflows in collaborative work of agents: humans, robots, and software. When cyber-physical systems emerge, new requirements and capabilities also emerge. Cyber-Collaborative Protocol (CCP) is a special type of TAPs that also administer and manage collaborative work, but are designed to do so for cyber-physical systems. For effective, automated and optimized cyber-physical operations, TAP and CCP are required to handle real-time complexities that are beyond communication, workflow, and coordination. TAP and CCP are introduced and defined in this chapter, along with examples of their design and application. Furthermore, current research in various production tasks utilizing TAP and CCP is summarized. In addition, the comparison of TAP and CCP with communication protocols and algorithms is explained and clarified. And just as better conductors will lead to better quality concerts, and superior operating systems will result in better computing, so do better designed TAPs and CCPs. This chapter aims to present the fundamental knowledge of TAP and CCP, which are the critical tools whose better design and effective utilization are the subject of this book.

1.1 Overview of Task Administration Protocols (TAPs) and Cyber-Collaborative Protocols (CCPs)

This section provides background knowledge about Task Administration Protocol (TAP) and Cyber-Collaborative Protocol (CCP). Also, this section explains TAP's/ CCP's roles in smart, optimized interactions, and in Cyber-Physical Systems (CPS).

P. O. Dusadeerungsikul and S. Y. Nof, *Cyber-Collaborative Algorithms and Protocols*, Automation, Collaboration, & E-Services 15,
https://doi.org/10.1007/978-3-031-56196-2_1

1.1.1 What is TAP, and Why is TAP needed?

To administer means to manage or control the operation of an organization, a business, or a system. In the modern world, the complexity of production and service systems increases significantly due to the integrations, communications, and interactions among participating system agents, such as resources, machines, human and robot workers and managers. A system such as service system, manufacturing system, warehouse system, and agricultural system comprises advanced technologies, massive flows of information, and multiple agents, intending to achieve higher levels of performance. The system, however, has a new challenge to overcome: more complex and timely interactions among internal and external agents (e.g., customer agents), sometimes under changes and disruptions. New technologies that enable decentralized and distributed agents drive the need for a methodology to collaborate, synchronize, and coordinate agents and their roles. Particularly, replanning, rescheduling, resequencing, reallocating, and reprioritizing become necessary. Therefore, a protocol called Task Administration Protocol (TAP) was developed for management and control of effective operations and interactions in systems [29, 42, 60, 65]. TAPs aim to optimize system performance by assigning and reassigning, synchronizing and resynchronizing, as needed, system agents and other system components. Each TAP utilizes rules, decision tools, heuristics, and interaction procedures managing agents, algorithms, and databases to achieve optimal workflow performance in dynamic and uncertain environments.

Figure 1.1 illustrates the general TAP model. Let us assume a system that can be decomposed to multiple operational stages, where a following stage is the consequence from the previous stage. To progress to the next stage, a decision needs to be made, and actions need to be executed. Furthermore, the system's progress keeps continuing until the system has reached the final stage in accomplishing its goal. Then, it repeats to achieve the next goal, again, with an optimized set of control, administration, or management decisions.

There will be a group of autonomous initiator agents ($\pi_i \subset \pi$) and autonomous responder agents ($\pi_r \subset \pi$) to decide and execute the operation at each stage. A group of agents will perform task ($\tau_a \subset \tau$) to obtain the transition stage ($\gamma_s \subset \gamma$). (Note that π is a set of participants or agents, τ is a set of problems or tasks, and γ is a set of transition stages).

Moreover, there will be a set of activities of each agent (φ). Note that $\varphi_k \subset \varphi$ will indicate the availability of the agent at a particular time. Therefore, the combination of π_i, τ_a, and φ_k will indicate the task dependency (δ_{ijk}) where δ is a set of dependency and $\delta_{ijk} \subset \delta$.

TAP, which applies a set of coordination/workflow protocols (σ), is designed, defined, and evaluated as a cyber tool for basic task-solving problems in a given system, under dynamic and uncertain changes and disruptions. TAP, as an advanced version of coordination protocol, will coordinate the agents to perform activities, with a coordination mechanism ($\sigma_p \subset \sigma$). A coordination protocol (σ_p) will harmonize, namely, coordinate, manage, synchronize, and prioritize δ_{ijk} so that the system

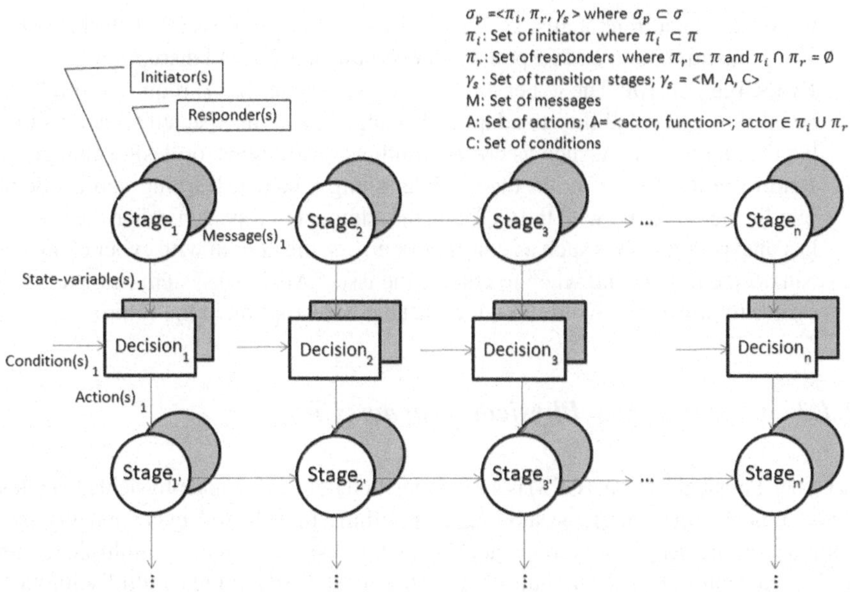

Fig. 1.1 General TAP model; able to handle, repeatedly, changing and disrupted conditions and state variables

achieves its goal, or series of goals, under the current, changing and disrupted conditions.

1.1.2 What is Cyber-Collaborative Protocol (CCP)?

CCP is a specific type of TAP, also designed for a decentralized and distributed system, but where the system is cyber augmented, or Cyber-Physical System. In a decentralized and distributed system, agents are given the authority to decide without command from the center (a centralized controller). Over time, there will be multiple requests and flows of information between agents. While a TAP itself can be cyber augmented (but meant for non-Cyber-Physical Systems), CCP has been derived from TAP with the augmentation of cyber element/method in the Cyber-Physical System (CPS) to enable smart interactions and collaborate interactions among agents. In this context, the cyber element/method in CPS comprises four aspects: computing, communication, real-time control, and brain models [63].

1. **Computing**: The cyber element of the system must be able to compute the given decisions about each task. Therefore, the computational power can be relatively larger or smaller due to the potential task assigned to the agents.
2. **Communication**: The communication element of cyber enables agents to connect and exchange information with each other. Communication allows agents

to update the current status, such as the location or progress of the task. It also allows identification and response to uncertain changes and disruptions.
3. **Real-time control**: The cyber element of the system that is required to handle a real-time control ability to adapt and change due to unexpected situations that happen in the system, such as errors, conflicts, disruptions, or design changes.
4. **Brain model**: With a brain model, integrating machine learning and artificial intelligence capacity with the CCP and the administered system, when the system has unexpected issues such as conflicts or errors, the system with cyber elements can decide if, how and when to resolve the issue. Also, the system can learn and evolve to prevent or avoid the same mistake or disruption in the future.

1.1.3 CCP in Cyber-Physical Systems (CPS)

A Cyber-Physical System (CPS) is a system which links cyber and physical attributes (capabilities) with computer systems and algorithms. In CPS, both cyber and physical components are deeply interconnected so that the system can receive multiple inputs and operate multiple tasks in multiple locations [98]. CPS requires multidisciplinary approaches (i.e., cybernetics, mechatronics, design, and process science) to achieve an effective system [91]. Also, CPS has an embedded system or control system, which tends to be based on computational and networked elements. CPS has shared the basic architecture of the Internet of Things/Internet of Services (IoT/IoS), but CPS represents a higher collaboration between cyber and physical components [69].

With this advancement of systems and infrastructure, CPS has enabled the cyber elements in TAPs to the level now called CCP. In order to achieve an effective CPS, CCP is an essential element. CCP plays a key role as it facilitates information flows and optimizes agents' interactions under dynamic and uncertain conditions. Besides, the CCP can present and resolve errors and conflicts in the system to assure and improve system performance and resilience.

An example of CCP in CPS is Automated Guided Vehicles (AGVs) in a smart factory. The AGV is responsible for moving packages or parts to their destination. Therefore, the AGV should be able to estimate the arrival time to the assigned destinations. Also, it can send a signal of the current location to update the current status. In some cases, the AGV needs to detour the current route according to an unexpected delay situation, such as congestion. When the planning algorithm is trained, it is expected that the algorithm will provide a better route for the AGV to minimize the travel time and cost, and eventually, lead to better system performance. All of the system elements, computational results, and communication among agents are coordinated by CCP. In addition, Table 1.1 indicates the cyber augmented parts in the AGV system.

In this example, we can see that the cyber elements that are also interconnected. For example, without the real-time communication that shares the current situation of the AGV, it is hard to calculate and inform about the arrival time, or dynamically adjust the route of the AGV. Besides, the brain model of the AGV requires computational power and memory to store and calculate the following, adjusted route.

Table 1.1 AGV in a factory: CPS systems

Features	Mapping cyber elements
Estimation of the arrival time	Computing
Sending a signal of the current location	Communication
Adjust route according to the current situation	Real-time control
Evolving based on learning the better route	Brain model (AI learning controller)

Recently, researchers have applied CCP in CPS in various contexts such as factories of the future, warehousing system, precision agriculture, etc. One of the prominent examples is the research from Guo and colleagues [20] that applied CCP workflow for agricultural CPS. They developed a protocol aiming at early detection of stress in greenhouse crop plants. The system has multiple agents (human, robot, sensors) as well as algorithms. Then, the designed system with CCP is compared with the current practice (non-CCP system). The results show that the designed cyber augmented system (CPS with CCP) outperforms the current practice. (More detail is presented in Chap. 3.)

1.1.4 Roles of TAP/CCP in Smart Interactions and Optimized Collaborations

For a decentralized and distributed system, TAP/CCP play two key roles: 1) Coordination network; and 2) Active middleware. In both roles, TAP/CCP have been designed when communication protocols, workflow protocols, and coordination protocols fell short in handling complex automation systems with uncertainties, dynamic changes, errors, conflicts, and disruptions.

1. Coordination Network:

In a system with multiple tasks and/or multiple agents, one crucial role of TAP/CCP is to establish control rules (logic) that enable effective collaboration among agents and tasks. It is expected that better collaboration will potentially lead to smarter interaction of the system both with other systems, and inside the system. With better automated collaboration, the system can have a higher potential to achieve its objectives (i.e., minimizing cost, time, defects, etc.). In this role, we can name TAP/CCP as a *regular TAP/CCP*. Figure 1.2 illustrates the regular TAP/CCP model.

2. **Active Middleware**:

Another critical role of TAP/CCP is to trigger and initiate interactions to optimize agent/system performance. For example, stop or enable the agent to work at the right time; negotiate or adjust rewards of the agent in the system; assign or allocate a given agent to the given task. This is an *Active TAP/CCP role*. Figure 1.3 illustrates TAP/CCP in the role of active middleware.

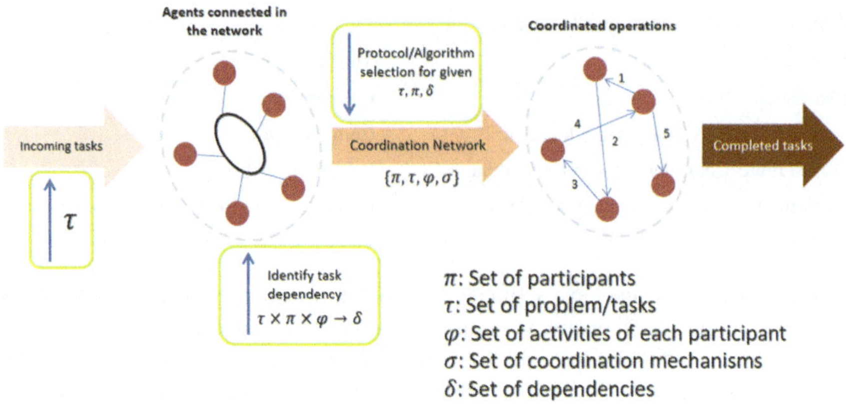

Fig. 1.2 TAP/CCP in the Role of an Advanced Coordination Network

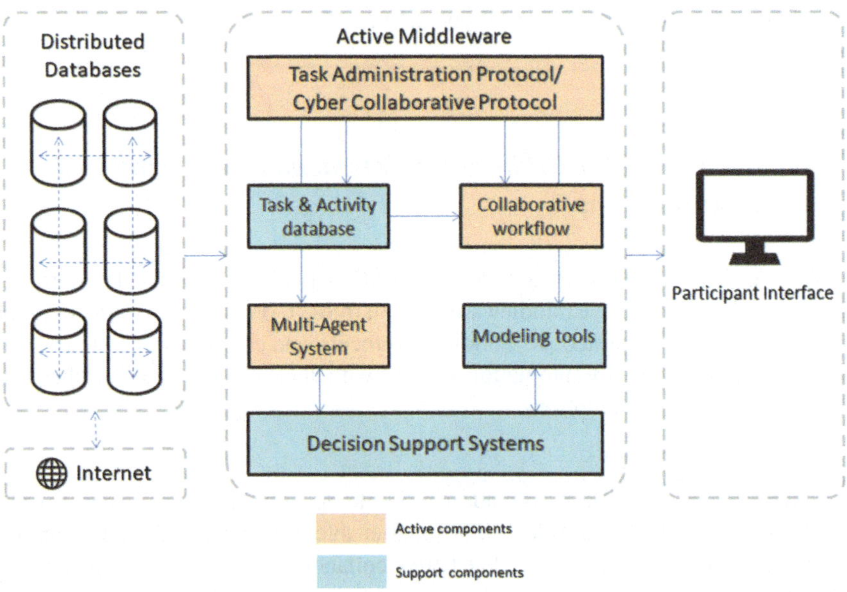

Fig. 1.3 TAP/CCP in the Role of Active Middleware

Engineers and researchers have utilized TAP/CCP to design effective workflow in various fields such as workflow design, task allocation design, timeout design, Petri-net model design, and Precision Agriculture (PA) [1, 15, 20, 34, 65]. The system with a TAP/CCP framework outperforms a non-TAP/non-CCP system as the activities in the system are synchronized [33]. Table 1.2 provides examples of TAP/CCP in different scenarios, a summarization of the results, and TAP's/CCP's roles in each scenario.

Table 1.2 Examples of TAP/CCP related articles

Production task	Summary	Roles of TAP/ CCP	References
Protocol design	The researchers/developers utilize TAP/ CCP to handle the complex environment. Tasks include initialization, allocation, and monitoring of the assignment. TAP/CCP is used for giving priority, reporting status, and monitoring the current task. The results show that the significant improvement of performance is due to a higher level of collaborative intelligence gained by the TAP/CCP	Advanced coordination network	[34]
Collaborative intelligence	The research presents a framework of e-Factories collaboration networks. The framework is augmented by Collaborative Control Theory (CCT). TAP/CCP is utilized for a better task-resource allocation model in the e-Factories		[53]
Multi-machine/robot coordination & task assignment/allocation	The research focuses on a multi-sensor task allocation framework for supply network security to maximize the number of security events that are correctly detected. TAP/CCP is applied in the monitoring layer for dealing with time-consuming tasks, conflicts in the monitoring process, and sensor task assignment. TAP/CCP has increased system performance in various scenarios, such as the situation that sensors produce conflicting signals		[96]
Demand and capacity sharing	A TAP/CCP is designed to allocate and re-allocate resources based on the priority of the demand in real-time. TAP/CCP, which enables the demand and capacity sharing process, improves a Predictive Best Matching Protocol	Active middleware	[53]
Share resources; resource allocation	The protocol is developed for optimizing and managing system agents, which are distributed and decentralized. To achieve the goals, agents need to adjust their own decision based on feedback from the distributed collaboration network		[1]

(continued)

Table 1.2 (continued)

Production task	Summary	Roles of TAP/ CCP	References
Time-out	The research aims at the analysis of the Time-out protocol, which is derived from TAP/CCP. The Time-out protocol can improve system performance by optimizing shared resource-time and equipment		[68]
Smart robotics in warehouse, greenhouse	The project emphasizes a monitoring system detecting crop plant stresses using Collaborative Control Protocol. The protocol derived from Collaborative Control Theory and TAP/CCP synchronizes system agents (human, mobile robot, and sensors) and applies algorithms to achieve system objectives. The experimental results show that the designed protocol can significantly improve system performance and reduce operating costs		[15]
Share resources; resource allocation	The research presents a computer-integrated method to increase throughput and reduce the inspection time of assembly via a client/server production configuration to achieve cost-effective distributed production facilities. TAPs/CCPs have been applied in three task administration phases (1) task prioritizing, (2) prioritized time limitation, and (3) resource allocation and reallocation. A wire harness assembly implementation case with computer integrated testers is utilized to test the protocol; results show the significant superiority of the newly designed protocol		[103]

1.2 TAP/CCP Applications and Implementations

Researchers and practitioners have utilized TAP/CCP in various fields to improve coordination and collaboration among multiple system agents in real collaborative automation systems. This section provides useful examples of TAP/CCP models and their design. The examples are summarized in Table 1.3, categorized based on production tasks and implemented application domains. Next, in Sect. 1.8 the full references of these examples of TAP/CCP projects are provided.

Table 1.3 Examples of TAP/CCP implementations in various production tasks

Production task	TAP/CCP	Example references
Multi-machine/robot coordination & task assignment/allocation	CCC: Collaboration, Coordination, and Control	[67]
	Integration of Computer Integrated Manufacturing (CIM) information	[30–32]
	Coordination in distributed manufacturing	[5]
	Multi-robots and sensors task assignment	[109]
		[96, 97]
Time-out	TOP: Time-Out Protocol/ Multi-agents	[2]
	TOP for single server	[68]
	TOP: Energy and latency considerations	[25, 27]
Share resources; resource allocation	Client–server integration of testers	[19]
	TestLAN	[103]
	Adaptive TestLAN	[104]
	Cooperative autonomous agents	[1]
	Tool sharing and best matching	[48]
CRP: collaboration requirement planning	CRP for multi-processors for assembly	[70, 71]
	CRP & Conflict Resolution	[66]
	Asynchronous CRP with configurability	[114]
EWP: e-Work parallelism	Integration of parallel design tasks	[29]
	FDL: Facility Description Language	[37, 39]
	Parallelism in distributed organizations	[6, 7]
	EWP in supply decision network	[82]
	Parallelism of pick & place operations	[50]

(continued)

Table 1.3 (continued)

Production task	TAP/CCP	Example references
ECDP: error & conflict detection & prevention	Error detection & prediction algorithms	[8, 10–12]
	Conflict resolution in collaborative design	[99]
	Conflict resolution in supply chain security	[38]
	Decentralized ECDP	[9]
	Conflict prevention in taxiway & runway	[36]
FTT: fault tolerance by teaming	FTTP: Fault-Tolerant Time-Out Protocol	[40, 41]
	Timeout Information Forwarding	[26, 27]
	Resilience by teaming in supply networks	[72, 73]
A/DN: associate/dissociate with a network	Join/Leave/Remain in a network	[13]
	Affiliate/dissociate for DCS in network	[107]
	Coalition formation for dynamic DSC	[87]
Best matching	BMP: Best Match Protocol for suppliers	[100, 101]
	Best Match Protocol for assembly	[102]
	BMP with interdependent preferences	[47]
	DCS & BMP in enterprise collaboration	[45]
DCS: demand and capacity sharing	DCS for enterprise networks	[106, 107]
	Capacity sharing for sustainability & balance	[87]
	Real-time optimization of DCS	[51]
Production harmonization	Collaborative production line control	[75]

(continued)

Table 1.3 (continued)

Production task	TAP/CCP	Example references
	Facility sensor network data interference	[24]
	Knowledge based service planning	[14]
	Sustainable wind energy distribution	[23]
CI: collaborative intelligence	CI in internetworked e-Work	[117]
	The collaborative factory of the future	[53]
	Co-Insight for collaborative DSS	[116]
	CCT and DSS	[62]
DLOC: dynamic lines of collaboration	Collaborative disruption response in CPS	[112]
	Resilience informatics	[56]
	CRDP: Collaboration in Response to Disruption Propagation	[57, 58]
Supply chain negotiations/supplier selection	Direct/indirect delivery negotiations	[82]
	Supplier selection by information sharing	[83, 89]
	BMP in order fulfillment	[4]
	Resource distribution in supply chain	[81]
	Service-component selection in cloud manufacturing	[54]
Smart robotics in warehouse, greenhouse	Differentiated service policy	[21]
	Agricultural robot routing	[15]
	Human-in-the-Loop in CPS	[90]
Protocol design	TAP/CCP design for collaborative production	[33, 34]

1.2.1 Task Administration Protocols/Cyber-Collaborative Protocols (Samples)

Tables 1.4a through 1.4p present a categorization of articles from journals, books, and conference proceedings from the recent decades, organized by their related fields.

Table 1.4a Sample of TAPs/CCPs in Multi-machine/robot coordination & task assignment/allocation

Publications	References
A. Multi-Machine/Robot Coordination & Task Assignment/Allocation	
Nof, S. Y., & Weill, R. (1992). Collaborative coordination control (CCC) of distributed Multi-Machine Manufacturing. (invited) CIRP Annals, 41(1), 441–445.	[67]
Ceroni, J. A., Matsui, M., & Nof, S. Y. (1999). Communication-Based Coordination Modeling in Distributed Manufacturing Systems. International Journal of Production Economics, 60, 281–287.	[5]
Kim, C. O., & Nof, S. Y. (1997). Coordination and Integration Models for Distributed and Heterogeneous CIM Information. Knowledge-Based Systems: Advanced Concepts, Techniques and Applications (587–601).	[30]
Kim, C. O., & Nof, S. Y. (2000). Investigation of PVM for the Emulation and Simulation of a Distributed CIM Workflow System. International Journal of Computer Integrated Manufacturing, 13(5), 401–409.	[31]
Kim, C. O., & Nof, S. Y. (2001). Design of collaboration framework for distributed CIM data activities. IIE Transactions, 33(7), 535–546.	[32]
Tkach, I., Edan, Y., Jevtić, A., & Nof, S. Y. (2013, October). Automatic Multi-Sensor Task Allocation Using Modified Distributed Bees Algorithm. Proceedings of 2013 IEEE International Conference on Systems, Man, and Cybernetics (1401–1406).	[92]
Zhang, L., Zhong, H., & Nof, S. Y. (2015). Adaptive Fuzzy Collaborative Task Assignment for Heterogeneous Multirobot Systems. International Journal of Intelligent Systems, 30(6), 731–762.	[109]
Tkach, I., Edan, Y., & Nof, S. Y. (2017). Multi-Sensor Task Allocation Framework for Supply Networks Security Using Task Administration Protocols. International Journal of Production Research, 55(18), 5202–5224.	[96]
Tkach, I., Jevtić, A., Nof, S. Y., & Edan, Y. (2018). A Modified Distributed Bees Algorithm for Multi-Sensor Task Allocation. Sensors, 18(3), 759–774.	[97]

Table 1.4b Sample of TAPs/CCPs in TOP: Time-Out Protocols

Publications	References
B. TOP: Time-Out Protocols	
Anussornnitisarn, P., Peralta, J., & Nof, S. Y. (2002). Time-out Protocol for Task Allocation in Multi-Agent Systems. Journal of Intelligent Manufacturing, 13(6), 511–522.	[2]
Peralta, J., Anussornnitisarn, P., & Nof, S. Y. (2003). Analysis of a Time-out Protocol and its Applications in a Single Server Environment. International Journal of Computer Integrated Manufacturing, 16(1), 1–13.	[68]
Jeong, W., & Nof, S. Y. (2007). Timeout-based Information Forwarding Protocol for Wireless Sensor Networks. International Journal of Distributed Sensor Networks, 3(4), 331–346.	[25]
Jeong, W., & Nof, S. Y. (2009). Design of Timeout-based Wireless Microsensor Network Protocols: Energy and Latency Considerations. International Journal of Sensor Networks, 5(3), 142–152.	[28]

Table 1.4c Sample of TAPs/CCPs in shared resources; resource allocation

Publications	References
C. Shared Resources; Resource Allocation	
Esfarjani, K., & Nof, S. Y. (1998). A Client-Server Model for Integration and Collaboration in Production Testing. International Journal of Production Research, 36(12), 3295–3321.	[19]
Williams, N. P., Liu, Y., & Nof, S. Y. (2002). TestLAN Approach and Protocols for the Integration of Distributed Assembly and Test Networks. International Journal of Production Research, 40(17), 4505–4522.	[103]
Williams, N. P., Liu, Y., & Nof, S. Y. (2003). Analysis of Workflow Protocol Adaptability in TestLAN Production Systems. IIE Transactions, 35(10), 965–972.	[104]
Anussornnitisarn, P., Nof, S. Y., & Etzion, O. (2005). Decentralized Control of Cooperative and Autonomous Agents for Solving the Distributed Resource Allocation Problem. International Journal of Production Economics, 98(2), 114–128.	[1]
Moghaddam, M., & Nof, S. Y. (2013, May). Dynamic Tool Sharing with Best Matching Protocols for Efficient Assembly Line Balancing. Proceedings of IFAC Intelligent Manufacturing Systems, Sao Paolo, Brazil.	[43]
Moghaddam, M., & Nof, S. Y. (2015). Balanceable Assembly Lines with Dynamic Tool Sharing and Best Matching Decisions—A Collaborative Assembly Framework. IIE Transactions, 47(12), 1363–1378.	[46]
Moghaddam, M., & Nof, S. Y. (2015). Real-time Administration of Tool Sharing and Best Matching to Enhance Assembly Lines Balanceability and Flexibility. Mechatronics, 31, 147–157.	[48]

Table 1.4d Sample of TAPs/CCPs in CRP: Collaboration Requirement Planning

Publications	References
D. CRP: Collaboration Requirement Planning	
Rajan, V. N., & Nof, S. Y. (1996). Cooperation Requirements Planning (CRP) for Multiprocessors: Optimal Assignment and Execution Planning. Journal of Intelligent and Robotic Systems, 15(4), 419–435.	[70]
Rajan, V. N., & Nof, S. Y. (1996). Minimal Precedence Constraints for Integrated Assembly and Execution Planning. IEEE Transactions on Robotics and Automation, Special Issue on Assembly and Task Planning, 12(2), 175–186.	[71]
Nof, S. Y., & Chen, J. (2003). Assembly and Disassembly: An Overview and Framework for Cooperation Requirement Planning with Conflict Resolution. Journal of Intelligent and Robotic Systems, 37(3), 307–320.	[66]
Scavarda, M., & Nof, S. Y. (2011, May). Cooperation Requirement Planning in Supply Decisions Networks—the Physical Dimension. Proceedings of IIE Annual Conference, Reno, Nevada.	[79]
Zhong, H., Nof, S. Y., & Berman, S. (2015). Asynchronous Cooperation Requirement Planning with Reconfigurable End-Effectors. Robotics and Computer-Integrated Manufacturing, 34, 95–104.	[114]

Table 1.4e Sample of TAPs/CCPs in EWP: e-Work Parallelism

Publications	References
E. EWP: e-Work Parallelism	
Khanna, N., Fortes, J. A., & Nof, S. Y. (1998). A Formalism to Structure and Parallelize the Integration of Cooperative Engineering Design Tasks. IIE Transactions on Design and Manufacturing, 30(1), 1–15.	[29]
Lara Gracia, M. A., Witzerman, J. P., & Nof, S. Y. (2000). Facility Description Language for Integrating Distributed Designs. International Journal of Production Research, 38(11), 2471–2488.	[39]
Ceroni, J. A., & Nof, S. Y. (2002). A Workflow Model Based on Parallelism for Distributed Organizations. Journal of intelligent Manufacturing, 13(6), 439–461.	[6]
Lara Gracia, M. A., & Nof, S. Y. (2003). Computer-Supported Conflict Resolution for Collaborative Facility Designers. International Journal of Production Research, 41(2), 207–233.	[37]
Ceroni, J. A., & Nof, S. Y. (2005). Task Parallelism in Distributed Supply Organizations: A Case Study in the Shoe Industry. Production Planning & Control, 16(5), 500–513.	[7]
Moghaddam, M., & Nof, S. Y. (2016). Parallelism of Pick-and-Place Operations by Multi-Gripper Robotic Arms. Robotics and Computer-Integrated Manufacturing, 42, 135–146.	[50]
Scavarda, M., Reyes Levalle, R., Lee, S., & Nof, S. Y. (2017). Collaborative e-Work Parallelism in Supply Decisions Networks: The Chemical Dimension. Journal of Intelligent Manufacturing, 28(6), 1337–1355.	[80]

Table 1.4f Sample of TAPs/CCPs in ECDP: Error & Conflict Detection & Prevention

Publications	References
F. ECDP: Error & Conflict Detection & Prevention	
Chen, X. W., & Nof, S. Y. (2007). Error Detection and Prediction Algorithms: Application in Robotics. Journal of Intelligent and Robotic Systems, 48(2), 225–252.	[8]
Velásquez, J. D., Lara Gracia, M. A., & Nof, S. Y. (2008). Systematic Resolution of Conflict Situations in Collaborative Facility Design. International Journal of Production Economics, 116(1), 139–153.	[99]
Lara Gracia, M. A., & Nof, S. Y. (2009). Conflict Resolution in Supply Chain Security. International Journal of Value Chain Management, 3(2), 168–186.	[38]
Chen, X. W., & Nof, S. Y. (2010). A Decentralised Conflict and Error Detection and Prediction Model. International Journal of Production Research, 48(16), 4829–4843.	[9]
Ko, H. S., Yoon, S. W., & Nof, S. Y. (2011, August). Intelligent Alert Systems for Error and Conflict Detection in Supply Networks. Proceedings of IFAC World Congress, Milan, Italy.	[35]
Chen, X. W., & Nof, S. Y. (2012). Agent-Based Error Prevention Algorithms. Expert Systems with Applications, 39(1), 280–287.	[10]
Chen, X. W., & Nof, S. Y. (2012). Conflict and Error Prevention and Detection in Complex Networks. Automatica, 48(5), 770–778.	[11]
Chen, X. W., & Nof, S. Y. (2012). Constraint-Based Conflict and Error Management. Engineering Optimization, 44(7), 821–841.	[12]
Landry, S. J., Chen, X. W., & Nof, S. Y. (2013). Decision Support Methodology for Dynamic Taxiway and Runway Conflict Prevention. Decision Support Systems, 55(1), 165–174.	[36]

Table 1.4g Sample of TAPs/CCPs in FTT: Fault Tolerance by Teaming

Publications	References
G. FTT: Fault Tolerance by Teaming	
Liu, Y., & Nof, S. Y. (2004). Distributed Microflow Sensor Arrays and Networks (DMFSN/A): Design of Architectures and Communication Protocols. International Journal of Production Research, 42(15), 3101–3115.	[40]
Jeong, W., & Nof, S. Y. (2008). Performance Evaluation of Wireless Sensor Network Protocols for Industrial Applications. Journal of Intelligent Manufacturing, 19(3), 335–345.	[26]
Liu, Y., & Nof, S. Y. (2008). Fault-Tolerant Sensor Integration for Micro Flow-Sensor Arrays and Networks. Computers & Industrial Engineering, 54(3), 634–647.	[41]
Jeong, W., & Nof, S. Y. (2009). A Collaborative Sensor Network Middleware for Automated Production Systems. Special issue on collaborative e-Work networks in industrial engineering, Computers & Industrial Engineering, 57(1), 106–113.	[27]
Tkach, I., Nof, S. Y., & Edan, Y. (2013, August). Fault Tolerant Task Allocation in a Multisensory Supply Chain Monitoring System. Proceedings of ICPR22, Iguacu Falls, Brazil.	[95]
Reyes Levalle, R., & Nof, S. Y. (2015). A Resilience by Teaming Framework for Collaborative Supply Networks. Computers & Industrial Engineering, 90, 67–85.	[72]
Reyes Levalle, R., & Nof, S. Y. (2015). Resilience by Teaming in Supply Network Formation and Re-Configuration. International Journal of Production Economics, 160, 80–93.	[73]

Table 1.4h Sample of TAPs/CCPs in A/DN: Associate/Dissociate with a Network

Publications	References
H. A/DN: Associate/Dissociate with a Network	
Chituc, C. M., & Nof, S. Y. (2007). The Join/Leave/Remain (JLR) Decision in Collaborative Networked Organizations. Computers & Industrial Engineering, 53(1), 173–195.	[13]
Yoon, S. W., & Nof, S. Y. (2011). Affiliation/Dissociation Decision Models in Demand and Capacity Sharing collaborative Network. International Journal of Production Economics, 130(2), 135–143.	[107]
Scavarda, M., Medel, R., Lee, S., & Nof, S. Y. (2012, May). A Market-Based Approach to Adaptive Reconfigurable Supply Networks. Proceedings of IIE Annual Conference, Orlando, Florida.	[78]
Seok, H., & Nof, S. Y. (2014). Dynamic Coalition Reformation for Adaptive Demand and Capacity Sharing. International Journal of Production Economics, 147, 136–146.	[88]

Table 1.4i. Sample of TAPs/CCPs in Best Matching

Publications	References
I. Best Matching	
Velásquez, J. D., & Nof, S. Y. (2008). A Best-Matching Protocol for Collaborative e-Work and e-Manufacturing. International Journal of Computer Integrated Manufacturing, 21(8), 943–956.	[100]
Velásquez, J. D., & Nof, S. Y. (2008). Integration of Machine-Vision Inspection Information for Best-Matching of Distributed Components and Suppliers. Computers in Industry, 59(1), 69–81.	[101]
Velásquez, J. D., & Nof, S. Y. (2009). Best-Matching Protocols for Assembly in e-Work Networks. (invited) Special Issue on Transport Logistics and Physical Distribution, International Journal of Production Economics, 122(1), 508–516.	[102]
Moghaddam, M., & Nof, S. Y. (2014). Combined Demand and Capacity sharing with Best Matching Decisions in Enterprise Collaboration. International Journal of Production Economics, 148, 93–109.	[45]
Moghaddam, M., & Nof, S. Y. (2015). Best-Matching with Interdependent Preferences—Implications for Capacitated Cluster Formation and Evolution. Decision Support Systems, 79, 125–137.	[47]
Moghaddam, M., & Nof, S. Y. (2015, May). Interdependent Preferences-Definitions and Impacts on Team Formation. Proceedings of IIE Annual Conference, Nashville, Tennessee.	[49]
Moghaddam, M., & Nof, S. Y. (2017). Best Matching Theory & Applications. Springer International Publishing. Automation, Collaboration, and E-Services (ACES) Book Series.	[52]

Table 1.4j Sample of TAPs/CCPs in DCS: Demand and Capacity Sharing

Publications	References
J. DCS: Demand and Capacity Sharing	
Yoon, S. W., & Nof, S. Y. (2009, May). Distributed Decision Protocols for Optimal Demand and Capacity Sharing in Collaborative Enterprise Networks. Proceeding of IIE Annual Conference, Miami, Florida.	[105]
Yoon, S. W., & Nof, S. Y. (2010). Demand and Capacity Sharing Decisions and Protocols in a Collaborative Network of Enterprises. Decision Support Systems, 49(4), 442–450.	[106]
Yoon, S. W., & Nof, S. Y. (2011). Cooperative Production Switchover Coordination for the Real-Time Order Acceptance Decision. International Journal of Production Research, 49(6), 1813–1826.	[108]
Seok, H.S., & Nof, S.Y. (2012, May). Demand and Capacity Negotiated Sharing by Two Manufacturers. Proceeding of IIE Annual Conference, Orlando, Florida.	[86]
Moghaddam, M. & Nof, S.Y. (2013, August) Best Matching and Task Administration Protocols for Effective Demand and Capacity Sharing. Proceedings of ICPR22, Iguassu Falls, Brazil. **Won Best Paper Award for Young Researchers.**	[44]
Seok, H., & Nof, S. Y. (2014). Collaborative Capacity Sharing Among Manufacturers on the Same Supply Network Horizontal Layer for Sustainable and Balanced Returns. International Journal of Production Research, 52(6), 1622–1643.	[87]
Moghaddam, M., & Nof, S. Y. (2016). Real-Time Optimization and Control Mechanisms for Collaborative Demand and Capacity Sharing. International Journal of Production Economics, 171, 495–506.	[51]

Table 1.4k Sample of TAPs/CCPs in Production Harmonization

Publications	References
K. Production Harmonization	
Seok, H., & Nof, S. Y. (2011, May). Decision Support Protocol for Sustainability Issues in Supply Networks. IIIE Annual Conference. Reno, Nevada.	[84]
Berman, S., & Nof, S. Y. (2011, August). Collaborative Control Theory for Robotic Systems with Reconfigurable End-Effectors. Proceedings of ICPR21, Stuttgart, Germany.	[3]
Tkach, I., Edan, Y., & Nof, S. Y. (2011, August). A Framework for Automatic Multi-Agents Collaboration in Target Recognition Tasks. Proceedings of ICPR21, Stuttgart, Germany.	[93]
Reyes Levalle, R., Scavarda, M., & Nof, S. Y. (2012). Collaborative Production Line Control for Collaborative Supply Networks. Proceedings of IFAC Symposium on Information Control Problems in Manufacturing, Bucharest, Romania.	[74]
Reyes Levalle, R., Scavarda, M., Seok, H., & Nof, S. Y. (2012). Protocols for Sustainable Collaboration in Supply Networks. Proceedings of International Conference on Modeling, Optimization & Simulation, Bordeaux, France.	[77]
Tkach, I., Edan, Y., & Nof, S. Y. (2012, May). Security of Supply Chains by Automatic Multi-Agents Collaboration. Proceedings of IFAC Symposium on Information Control Problems in Manufacturing, Bucharest, Romania.	[94]
Devadasan, P., Zhong, H., & Nof, S. Y. (2013). Collaborative Intelligence in Knowledge Based Service Planning. Expert Systems with Applications, 40(17), 6778–6787.	[14]
Reyes Levalle, R., Scavarda, M., & Nof, S. Y. (2013). Collaborative Production Line Control: Minimisation of Throughput Variability and WIP. (invited) International Journal of Production Research, 51(23–24), 7289–7307.	[75]
Reyes Levalle, R., Scavarda, M., & Nof, S. Y. (2013, May). Collaborative Management of Supply Network Operations. Proceedings of IIE Annual Conference, San Juan, Puerto Rico.	[76]
Nof, S. Y. (2013, July). Sustainability and Resiliency in Supply Networks. Proceedings of Asia Pacific Industrial Engineering and Management Society (APIEMS14).	[61]
Jeong, W., Ko, H. S., Lim, H., & Nof, S. Y. (2013). A Protocol for Processing Interfered Data in Facility Sensor Networks. International Journal of Advanced Manufacturing Technology, 67(9), 2377–2385.	[24]
Jahanpour, E., Ko, H. S., & Nof, S. Y. (2014). Collaboration Platform for Sustainable Wind Energy Distribution Network. Proceedings of IFAC World Congress, Cape Town, South Africa.	[22]
Jahanpour, E., Ko, H. S., & Nof, S. Y. (2016). Collaboration Protocols for Sustainable Wind Energy Distribution Networks. International Journal of Production Economics, 182, 496–507.	[23]

Table 1.41 Sample of TAPs/CCPs in CI: Collaborative Intelligence

Publications	References
L. CI: Collaborative Intelligence	
Seok, H., & Nof, S. Y. (2011, August). The HUB-CI Initiative for Cultural, Education and Training, and Healthcare Networks. Proceedings of ICPR21, Stuttgart, Germany.	[85]
Zhong, H., & Nof, S. (2013, August). Collaborative Design for Assembly: the HUB-CI Model. Proceedings of ICPR22, Iguassu Falls, Brazil.	[110]
Zhong, H., Wachs, J. P., & Nof, S. Y. (2013). HUB-CI Model for Collaborative Telerobotics in Manufacturing. Proceedings of IFAC Intelligent Manufacturing Systems, Sao Paolo, Brazil.	[118]
Zhong, H., Nof, S. Y., & Ozsoy, E. (2015, May). Co-Viz: Matching Tools in Collaborative Visual Analytics. Proceedings of IIE Annual Conference, Nashville, Tennessee.	[115]
Zhong, H., Reyes Levalle, R., Moghaddam, M., & Nof, S. Y. (2015). Collaborative Intelligence—Definitions and Measured Impacts on Internetworked e-Work. Management and Production Engineering Review, 6(1), 67–78.	[117]
Zhong, H., Ozsoy, E., & Nof, S. Y. (2016). Co-Insights Framework for Collaborative Decision Support and Tacit Knowledge Transfer. Expert Systems with Applications, 45, 85–96.	[116]
Moghaddam, M., & Nof, S. Y. (2017). The Collaborative Factory of the Future. International Journal of Computer Integrated Manufacturing, Special Issue on Factory of the Future, 30(1), 23–43.	[53]
Nof, S. Y. (2017). Collaborative Control Theory and Decision Support Systems. (invited), Computer Science Journal of Moldova, 25(2) , 115–144.	[62]
Dusadeerungsikul, P. O., Sreeram, M., He, X., Nair, A., Ramani, K., Quinn, A. J., & Nof, S. Y. (2019, August). Collaboration Requirement Planning Protocol for HUB-CI in Factories of the Future. Procedia manufacturing, ICPR-25, Chicago, Illinois.	[18]
Nof, S.Y. (2019, December). From Integration to Augmentation, from Interaction to Collaborative Control—IE/MS Frontiers for Future Work and Factories. Proceedings of APIEMS-IFPR_APR Conference, Kanazawa, Japan.	[64]

Table 1.4m Sample of TAPs/CCPs in DLOC: Dynamic Lines of Collaboration

Publications	References
M. DLOC: Dynamic Lines of Collaboration	
Zhong, H., & Nof, S. Y. (2014, July). DLOC complex network model for supply network disruption response. Proceedings of ICPR-AEM, Cluj-Napoca, Romania.	[111]
Zhong, H., & Nof, S. Y. (2015). The Dynamic Lines of Collaboration Model: Collaborative Disruption Response in Cyber–Physical Systems. Computers & Industrial Engineering, 87, 370–382.	[112]
Nguyen, W. P., & Nof, S. Y. (2018). Resilience Informatics for Cyber-augmented Manufacturing Networks (CMN): Centrality, Flow and Disruption. Studies in Informatics and Control, 27(4), 377–384.	[56]
Nguyen, W. P., Nair, A. S., & Nof, S. Y. (2019, August). Advancing Cyber-Physical Systems Resilience: The Effects of Evolving Disruptions. Procedia manufacturing, ICPR-25, Chicago, Illinois.	[55]
Nguyen, W. P., & Nof, S. Y. (2019). Collaborative Response to Disruption Propagation (CRDP) in Cyber-Physical Systems and Complex Networks. Decision Support Systems, 117, 1–13.	[57]
Nguyen, W. P., & Nof, S. Y. (2020). Strategic Lines of Collaboration in Response to Disruption Propagation (CRDP) Through Cyber-Physical Systems. International Journal of Production Economics, 230, 107865.	[58]
Zhong, H., & Nof, S. Y. (2020). Dynamic Lines of Collaboration: Disruption Handling & Control. Springer International Publishing. Automation, Collaboration, and E-Services (ACES) Book Series.	[113]

Table 1.4n Sample of TAPs/CCPs in Supply Chain Negotiations/Supplier Selection

Publications	References
N. Supply Chain Negotiations/Supplier Selection	
Scavarda, M., Seok, H., Puranik, A. S., & Nof, S. Y. (2015). Adaptive Direct/Indirect Delivery Decision Protocol by Collaborative Negotiation Among Manufacturers, Distributors, and Retailers. International Journal of Production Economics, 167, 232–245.	[82]
Bhargava, R., Reyes Levalle, R., & Nof, S. Y. (2016). A Best-Matching Protocol for Order Fulfillment in Re-Configurable Supply Networks. Computers in Industry, 82, 160–169.	[4]
Seok, H., Kim, K., & Nof, S. Y. (2016). Intelligent Contingent Multi-Sourcing Model for Resilient Supply Networks. Expert Systems with Applications, 51, 107–119.	[83]
Scavarda, M., Seok, H., & Nof, S. Y. (2017). The Constrained-Collaboration Algorithm for Intelligent Resource Distribution in Supply Networks. Computers & Industrial Engineering, 113, 803–818.	[81]
Moghaddam, M., & Nof, S. Y. (2018). Collaborative Service-Component Integration in Cloud Manufacturing. (invited) International Journal of Production Research, 56(1–2), 677–691.	[54]
Seok, H., & Nof, S. Y. (2018). Intelligent Information Sharing Among Manufacturers in Supply Networks: Supplier Selection Case. Journal of Intelligent Manufacturing, 29(5), 1097–1113.	[89]

Table 1.4o Sample of TAPs/CCPs in Smart Robotics in Warehouse, Greenhouse

Publications	References
O. Smart Robotics in Warehouse, Greenhouse	
He, Z., Aggarwal, V., & Nof, S. Y. (2018). Differentiated Service Policy in Smart Warehouse Automation. (invited) International Journal of Production Research, 56(22), 6956–6970.	[21]
Dusadeerungsikul, P. O., & Nof, S. Y. (2019). A Collaborative Control Protocol for Agricultural Robot Routing with Online Adaptation. Computers & Industrial Engineering, 135, 456–466.	[15]
Dusadeerungsikul, P. O., Nof, S. Y., Bechar, A., & Tao, Y. (2019, August). Collaborative Control Protocol for Agricultural Cyber-Physical System. Procedia Manufacturing, ICPR-25, Chicago, Illinois.	[17]
Sreeram, M., & Nof, S. Y. (2021). Human-In-The-Loop: Role in Cyber Physical Agricultural Systems. International Journal of Computers, Communications and Control, 16(2).	[90]
Dusadeerungsikul, P. O., Nof, S. Y. (2024). Precision agriculture with AI-based responsive monitoring algorithm. International Journal of Production Economics. 109204.	[16]

Table 1.4p Sample of TAPs/CCPs in TAP/CCP Protocol Design

Publications	References
P. TAP/CCP Protocol Design	
Nof, S. Y. (editor, co-author). (1994). Information and Collaboration Models of Integration, Kluwer Academic Publishers, Dordrecht, 480 pages.	[59]
Ko, H. S., & Nof, S. Y. (2010). Design of protocols for task administration in collaborative production systems. International Journal of Computers Communications & Control, 5(1), 91–105.	[33]
Ko, H. S., & Nof, S. Y. (2012). Design and Application of Task Administration Protocols for Collaborative Production and Service Systems. International Journal of Production Economics, 135(1), 177–189.	[34]
Nof, S. Y., Ceroni, J., Jeong, W., & Moghaddam, M. (2015). Revolutionizing Collaboration through e-Work, e-Business, and e-Service (Vol. 2). Springer Publishers, Automation, Collaboration, and E-Services (ACES) Book Series.	[65]

1.3 Comparing TAP/CCP with Communication Protocols and with Algorithms

1.3.1 Differences and Similarities of TAPs/CCPs and Communication Protocols

TAPs and CCPs are sometimes confused with communication protocols. Indeed, they are all protocols, therefore, they share some commonalities. But they are distinctly different, and have different objectives and roles. Overall, it must be stated that communication protocols are essential for communications between systems and among the system's participants, and reliable communication is essential for effective, collaborative task performance. But understandably, even good communication is not sufficient, for instance, it may not be timely, it may be incomplete, and once the information is communicated, how is it going to be used and benefit the agents/actors/participants? For decentralized and distributed agents to work effectively and harmoniously, platforms that facilitate, communicate, and collaborate all agents must be developed. Previously, we discussed TAP/CCP, which are advanced coordination and collaboration protocols. The main objective of TAP/CCP is to allocate and administer tasks, resources, and agents, given that communication between and among all participants is available and reliable.

A communication protocol is focused on how decentralized and distributed agents can communicate synchronously and asynchronously. This protocol is a mechanism with digital transmission and telecommunicating rules for information exchanges. Once the intended information has been received, the role of the communication protocol ends. The role of TAP/CCP begins.

While TAP/CCP and communication protocol share the objective of facilitating the system to have better performance, they are utilized as different means and for different roles. Table 1.5 compares between Communication protocols and TAPs/CCPs in general flow systems, production, service, and logistics.

Next, the purpose of the following section is to illustrate the comparison, using two examples in precision farming: wireless sensor network (WSN) and Decision Support System (DSS). Specific similarities and differences between communication protocols and TAPs/CCPs that have been identified as necessary for the implementation of cyber methods in these two examples are explained.

Table 1.6 below identifies two communication protocols, and two TAPs/CCPs that are required for the management of precision farming. Table1.7 describes the similarities between these communication protocols and TAPs/CCPs. Table 1.8 describes the differences between these communication protocols and TAPs/CCPs that have been identified.

Table 1.5 Comparison of communication protocols versus TAPs/CCPs

Design feature	Communication protocols	TAPs/CCPs
Purpose	The purpose of all standard communication protocols is the same: Perfectly transmitting a message from a sender to one or more receivers	The purpose of TAPs/CCPs depends on the specific application, and for a given application could be automatically adjusted according to the changing requirements and the current resources and conditions of the system
Design purpose	Standard communication protocols are designed for a specific hardware/software technology to enable communication between elements in a network, e.g., a Local Area Network (LAN) protocol cannot be directly implemented for Wireless Local Area Network (WLAN)	TAPs/CCPs are independent of the specific hardware/software technology of the systems, but specific protocols that are controlled by the TAP/CCP and parameter values must be determined for the application context, e.g., a TAP/CCP designed for best matching of a given set of suppliers to a set of clients
Logic techniques used by the protocols	The basic logic of standard communication protocols is common: (1) message encoding; (2) formatting and encapsulation; (3) timing; (4) message size; and (5) message delivery options	TAPs/CCPs do not have standard elements; while similar logic mechanisms are reusable, components of TAPs/CCPs and parameter values must be designed and refined for the specific application and context
Optimization	Five common optimization methods: (1) compression; (2) object pipelining; (3) transmission sequencing; (4) layering and event tracing; (5) parallelization	TAPs/CCPs optimize systems operations by triggering specific procedures or algorithms, depending on the current conditions of the system, e.g., implementation of a mixed integer algorithm to best-match currently available tools to requesting assembly robots
Interactions with agents/actors	Standard communication protocols are passive, in the sense that they do not initiate but are activated for interactions between communicating agents, e.g., they are activated only when an entity needs to transmit an information message to another entity (or entities) in a network	TAPs/CCPs are typically active and initiate interactions between agents depending on the conditions or states of the system, e.g., when a load is consolidated and ready at a distribution center, it triggers a TAP/CCP that requests data about incoming and available carrier trucks

(continued)

Table 1.5 (continued)

Design feature	Communication protocols	TAPs/CCPs
Layers of design	Layers of design are standard for all communication protocols: physical, data link, network, transport, session, presentation, and application	Four general application layers are used to implement TAPs/CCPs: (1) Work execution layer, e.g., workers, machines, computers; (2) digital/virtual/twin/simulation models of these work agents and their teams; (3) local and internal networks, e.g., Intranet, IoT/IoS, sensors; supervisory networks/Enterprise Resource Planning (ERP); (4) external networks, e.g., extranet, wide area networks
Performance metrics, QoS Indicators	Common Quality of Service indicators for standard communication protocols include latency, noise elimination, reliability, availability, and fidelity	Quality of Service indicators of TAPs/CCPs are specific for the application context, e.g., percentage of tasks completed per time interval; percentage of assembly errors per time interval; total empty driven miles per delivered container, or per AGV, probability of idle bottlenecks
Application example	Transmission Control Protocol/Internet Protocol (TCP/IP) for Intranet communication between tablets or cellphones of truck drivers and the main server at a cross-docking distribution center	TAP/CCP for schedule and truck-load assignment optimization, and conflict prevention in the distribution center application
Public Availability	Must be standard and publicly known	Competitive, proprietary; must be known only to the participating systems and agents

Table 1.6 Example of communication protocols and TAPs/CCPs identified

Communication protocols	TAPs/CCPs
Wireless Local Area Network (WLAN)—wireless Highway Addressable Remote Transducer (HART) Protocol	Optimized Cluster-Based Dynamic Energy-Aware Routing Protocol with Gateway Clustering Energy-Efficient Centroid-(GCEEC-) based routing protocol (GCEEC)
Wide-Area Network (WAN)—TCP/IP; WLAN—wireless HART Protocol	Decision Support System for Precision Farming

Table 1.7 Similarities between example communication protocols and TAPs/CCPs

Design Feature	Communication protocols	TAPs/CCPs
Purpose example	Required for setup and maintenance of networks. In a WLAN network, routers are in a constant state of information exchange to determine best routes and to avoid broadcast storms	Makes decisions regarding how the network should be setup. In the GCEEC, gateway and cluster head routers are determined by the base station to optimize energy usage
Optimization techniques example	Optimized through feedback from the overall network. For example, in the application of Transmission Control Protocol (TCP), acknowledgment packets sent from the receiver to the sender can alert the sender about congestion in the network allowing the sender to optimize sending patterns	Optimized through feedback from networks. In the DSS, an irrigation schedule is determined from data on the WAN but is optimized through feedback from the WSN (Wireless Sensor Network)
Performance metrics example	Performance evaluated based on reliability of data exchange. In a WAN network, the transport layer, such as TCP, is considered to be reliable if it resends all dropped or corrupted packets and if it attempts to make progress. In this scenario, attempting to make progress means if there is data to be sent, the transport layer will eventually attempt to send that data	Performance is evaluated based on the reliability of the decisions being made. For a DSS, the performance of the system can be evaluated by the reliability of the determined irrigation schedule, and whether the agents are able to execute the scheduled tasks
Interactions with agents example	Requires constant communication between agents to handle network failures and optimize routes. In a WLAN, control packets are exchanged to handle link failures	Requires constant communication between agents to make decisions about the higher-level network wireless topology. In the DSS example, the base station receives information from the WSN and updates its models accordingly
Logic techniques example	Applies a variety of algorithms to handle different issues. In the example of the WAN, algorithms are available to assist the protocol with flow and congestion control	Uses algorithms to assist in decision making. The DSS uses an algorithm based on historical weather data and crop plant parameters to recommend a suitable irrigation schedule
Adaptiveness example	Must be able to adapt to changes in the network and handle them without a noticeable loss in performance. The WAN is in a constant state of change and TCP is responsible for ensuring data is still transmitted despite these changes	Must be able to adapt its decisions based on incoming information and changing conditions. The GCEEC protocol will find new cluster heads based on the residual energy centroids within the WSN

Table 1.8 Differences between example communication protocols and TAPs/CCPs

Design feature	Communication protocols	TAPs/CCPs
Purpose example	Purpose is to ensure data is reliably transmitted between senders and receivers. In both communication protocols identified, the purpose is to ensure information is flowing correctly and reliably	The TAPs/CCPs are designed to be responsible for managing multiple tasks. In the case of the GCEEC protocol, the TAP/CCP is responsible for selecting cluster heads and gateway nodes, and ensuring the network is aware of the decisions
Optimization techniques example	Protocols are optimized to ensure stable communication. In TCP, optimization occurs based on bandwidth usage so as not to overload the network, which would result in loss of packets	In the example of the GCEEC TAP/CCP, the network is first optimized based on energy management and then the focus shifts to fastest path techniques
Performance metrics example	Communication protocols are only focused on delivering information reliably over networks that are not necessarily reliable, therefore performance is measured by the amount of uncorrupted data received	TAPs/CCPs have many responsibilities. In the DSS example, the performance can be measured by how well tasks are administered to agents regarding the irrigation scheduling execution
interactions with agents example	Interactions with agents are used to manage failures within networks. WLAN networks use control packets to develop fastest transmission to destinations	TAPs/CCPs can interact with agents to get information but must also command different agents to complete tasks. In the DSS example, the system interacts with the sensor nodes and data bases to generate irrigation schedules, and must also relay this information to the agents involved in executing the irrigation
Logic techniques example	Logic techniques are focused on ensuring smooth communication and overcoming failures within the network	Logic techniques are focused on optimizing resource allocation by properly managing tasks' and agents' interactions

1.3.2 *Differences and Similarities of TAP/CCP and Algorithms*

Commonly known, algorithms are the essential elements of a smart system, as they enable agents to decide on how to perform and deliver outputs to achieve an objective. Algorithms, however, are relatively static and not flexible to any dynamic changes in input, or in the environment of input, or the coordination needs with other decisions. On the other hand, protocols (i.e., TAP/CCP) are designed to be flexible to handle and respond to different inputs and coordination needs, as they are composed of multiple decision-making tools and procedures to select the appropriate algorithms for each given set of conditions. Often, multiple algorithms have to be activated.

Table 1.9 compares between Algorithms and TAPs/CCPs in general flow systems, production, service, and logistics.

Table 1.9 Comparison of TAPs/CCPs versus algorithms

Design feature	Algorithms	TAPs/CCPs
Design purpose	Algorithms are designed to generate an output by performing calculations that follow a specific logic based on a given set of inputs	TAPs/CCPs are designed to initiate interactions between agents and activate protocols depending on the conditions of the system to ensure a smooth execution of tasks in a dynamic and unpredictable environment
Logic	The order or sequence of instructions is important for an algorithm to work well. *For example:* if the algorithm is initiated from the middle part of its code, the quality of the output is compromised	TAPs/CCPs are sensitive to chronology or sequence of events. *For example:* Based on given events, the appropriate part of the TAP/CCP is activated first. For this reason, Petri Nets are a good modeling approach for TAPs/CCPs
Design complexity	Algorithms typically need to involve complex computational procedures to be effective. Usually, the larger the size of the problem, the higher the complexity of the algorithm	TAPs/CCPs typically do not need to involve complex procedures to be effective. If a simple decision logic is appropriate now, it will be implemented. The reverse is also true
Optimality	Algorithms allow proof of optimality: it can be determined if the result of the algorithm is optimal, or how far the solution is from optimal	Performance of different TAPs/CCPs may be compared based on quality of service measures, but optimality of a TAP/CCP may be unknown
Interactions with agents/actors	Algorithms do not interact directly with agents, they just provide an action recommendation when requested	TAPs/CCPs mediate interactions of agents and trigger procedure and algorithms to support collaboration in executing tasks in the system

(continued)

(continued)

Design feature	Algorithms	TAPs/CCPs
Activation	Parameter values (inputs) and a formal request to run the logic are required to activate an algorithm	Conditions or states of the system activate TAPs/CCPs
Performance metrics, QoS Indicators	Performance of an algorithm can be measured, after running it, by: (1) computational time, and (2) solution quality (relative to the optimal solution)	Performance of a TAP/CCP can be assessed by: (1) simulating the system under the management of the given TAP/CCP, or (2) activating and operating the system under the management of the given TAP/CCP

To illustrate, Tables 1.10, 1.11, and 1.12 compare algorithms and TAPs/CCPs specific to the precision farming example. Table 1.10 shows the algorithm and corresponding TAP/CCP identified for correct implementation of the cyber-method. Table 1.11 describes the similarities between the algorithms and TAPs/CCPs that have been identified. Table 1.12 describes the differences between the algorithms and TAPs/CCPs that have been identified for the precision farming example.

Table 1.10 Algorithms and TAPs/CCPs identified (example)

Algorithm	TAP/CCP
Resource Sharing for precision farming irrigation, computing the daily soil water balance, e.g., FAOS56	Real-time decision support systems for precision farming, planning/optimizing, operating, executing and controlling multiple resources, e.g., Demand and Capacity Sharing

Table 1.11 Similarities between Algorithms and TAPs/CCPs (examples)

Design Feature	Algorithm	TAPs/CCPs
Purpose	Determine the correct amount of water and nutrients for the crop plants while not over-utilizing any resources	Ensure the correct amount of water and nutrients are being delivered to the crop plants
Optimization techniques	Optimized by updating parameters and inputs so the model better matches the actual field. In the FAOS56, crop plant and weather parameters are updated as needed	TAPs/CCPs use algorithms to determine the best course of action, so as an algorithm is optimized, the corresponding TAP/CCP will also be optimized
Performance metrics	The algorithm's performance can be measured based on speed and closeness to the optimal result. In the FAOS6 algorithm, speed is imperative as the irrigation schedule may need to change quickly based on environmental factors	A performance metric for the DSS is also speed but the speed relates to the time it takes to administer tasks instead of the time it takes to determine the best results

(continued)

7. Ceroni JA, Nof SY (2005) Task parallelism in distributed supply organizations: a case study in the shoe industry. Prod Planning Control 16(5):500–513. https://doi.org/10.1080/09537280500121851
8. Chen XW, Nof SY (2007) Error detection and prediction algorithms: application in robotics. J Intell Rob Syst 48(2):225–252. https://doi.org/10.1007/s10846-006-9094-9
9. Chen XW, Nof SY (2010) A decentralised conflict and error detection and prediction model. Int J Prod Res 48(16):4829–4843. https://doi.org/10.1080/00207540903067201
10. Chen XW, Nof SY (2012) Agent-based error prevention algorithms. Expert Syst Appl 39(1):280–287. https://doi.org/10.1016/j.eswa.2011.07.018
11. Chen XW, Nof SY (2012) Conflict and error prevention and detection in complex networks. Automatica 48(5):770–778. https://doi.org/10.1016/j.automatica.2012.02.030
12. Chen XW, Nof SY (2012) Constraint-based conflict and error management. Eng Optim 44(7):821–841. https://doi.org/10.1080/0305215X.2011.613466
13. Chituc C-M, Nof SY (2007) The Join/Leave/Remain (JLR) decision in collaborative networked organizations. Comput Ind Eng 53(1):173–195. https://doi.org/10.1016/j.cie.2007.05.002
14. Devadasan P, Zhong H, Nof SY (2013) Collaborative intelligence in knowledge based service planning. Expert Syst Appl 40(17):6778–6787. https://doi.org/10.1016/j.eswa.2013.06.048
15. Dusadeerungsikul PO, Nof SY (2019) A collaborative control protocol for agricultural robot routing with online adaptation. Comput Indus Eng, 135.https://doi.org/10.1016/j.cie.2019.06.037
16. Dusadeerungsikul, P. O., Nof, S. Y. (2024). Precision agriculture with AI-based responsive monitoring algorithm. International Journal of Production Economics. 109204.
17. Dusadeerungsikul PO, Nof SY, Bechar A, Tao Y (2019) Collaborative control protocol for agricultural cyber-physical system. Procedia Manuf ICPR-25 39:235–242. https://doi.org/10.1016/j.promfg.2020.01.330
18. Dusadeerungsikul PO, Sreeram M, He X, Nair A, Ramani K, Quinn AJ, Nof S Y (2019) Collaboration requirement planning protocol for HUB-CI in factories of the future. Procedia Manuf ICPR-25 39:218–225. https://doi.org/10.1016/j.promfg.2020.01.327
19. Esfarjani K, Nof SY (1998) Client-server model of integrated production facilities. Int J Prod Res 36(12):3295–3321. https://doi.org/10.1080/002075498192076
20. Guo P, Dusadeerungsikul PO, Nof SY (2018) Agricultural cyber physical system collaboration for greenhouse stress management. Comput Electron Agric 150:439–454. https://doi.org/10.1016/j.compag.2018.05.022
21. He Z, Aggarwal V, Nof SY (2018) Differentiated service policy in smart warehouse automation. Int J Prod Res 56(22):6956–6970. https://doi.org/10.1080/00207543.2017.1421789
22. Jahanpour E, Ko HS, Nof SY (2014) Collaboration platform for sustainable wind energy distribution network. Proc IFAC World Congress 47(3):4266–4271. https://doi.org/10.3182/20140824-6-ZA-1003.01110
23. Jahanpour E, Ko HS, Nof SY (2016) Collaboration protocols for sustainable wind energy distribution networks. Int J Prod Econ 182:496–507. https://doi.org/10.1016/j.ijpe.2016.09.010
24. Jeong W, Ko HS, Lim H, Nof SY (2013) A protocol for processing interfered data in facility sensor networks. Int J Adv Manuf Technol 67(9–12):2377–2385. https://doi.org/10.1007/s00170-012-4657-3
25. Jeong W, Nof SY (2007) Timeout-based information forwarding protocol for wireless sensor networks. Int J Distrib Sens Netw 3(4):331–346. https://doi.org/10.1080/15501320601062148
26. Jeong W, Nof SY (2008) Performance evaluation of wireless sensor network protocols for industrial applications. J Intell Manuf 19(3):335–345. https://doi.org/10.1007/s10845-008-0086-4
27. Jeong W, Nof SY (2009) A collaborative sensor network middleware for automated production systems. Comput Ind Eng 57(1):106–113. https://doi.org/10.1016/j.cie.2008.11.007

28. Jeong W, Nof SY (2009) Design of timeout-based wireless microsensor network protocols: energy and latency considerations. Int J Sensor Netw 5(3):142. https://doi.org/10.1504/IJSNET.2009.026362
29. Khanna N, Fortes JAB, Nof SY (1998) A formalism to structure and parallelize the integration of cooperative engineering design tasks. IIE Trans 30(1):1–15. https://doi.org/10.1080/07408179808966433
30. Kim CO, Nof SY (1997) Coordination and integration models for distributed and heterogeneous CIM information. In: Tzafestas SG (ed) Knowledge Based Systems: Advanced Concepts, Techniques & Applications. World Scientific, pp 587–602
31. Kim CO, Nof SY (2000) Investigation of PVM for the emulation and simulation of a distributed CIM workflow system. Int J Comput Integr Manuf 13(5):401–409. https://doi.org/10.1080/09511920050117892
32. Kim CO, Nof SY (2001) Design of collaboration framework for distributed CIM data activities. IIE Trans 33(7):535–546. https://doi.org/10.1080/07408170108936852
33. Ko HS, Nof SY (2010) Design of protocols for task administration in collaborative production systems. Commun Control 5(1):91–105
34. Ko HS, Nof SY (2012) Design and application of task administration protocols for collaborative production and service systems. Int J Prod Econ 135(1):177–189. https://doi.org/10.1016/j.ijpe.2011.06.005
35. Ko HS, Yoon SW, Nof SY (2011) Intelligent alert systems for error and conflict detection in supply networks. Proc IFAC World Congress 44(1):1602–1607. https://doi.org/10.3182/20110828-6-IT-1002.03841
36. Landry SJ, Chen XW, Nof SY (2013) A decision support methodology for dynamic taxiway and runway conflict prevention. Decis Support Syst 55(1):165–174. https://doi.org/10.1016/j.dss.2013.01.016
37. Lara Gracia MA, Nof SY (2003) Computer-supported conflict resolution for collaborative facility designers. Int J Prod Res 41(2):207–233. https://doi.org/10.1080/0020754021000030411
38. Lara Gracia MA, Nof SY (2009) Conflict resolution in supply chain security. Int J Value Chain Manag 3(2):168. https://doi.org/10.1504/IJVCM.2009.026956
39. Lara Gracia MA, Witzerman JP, Nof SY (2000) Facility description language for integrating distributed designs. Int J Prod Res 38(11):2471–2488. https://doi.org/10.1080/00207540050031878
40. Liu Y, Nof SY (2004) Distributed microflow sensor arrays and networks: design of architectures and communication protocols. Int J Prod Res 42(15):3101–3115. https://doi.org/10.1080/00207540410001699363
41. Liu Y, Nof SY (2008) Fault-tolerant sensor integration for micro flow-sensor arrays and networks. Comput Ind Eng 54(3):634–647. https://doi.org/10.1016/j.cie.2007.09.013
42. Miranda PA, Garrido RA, Ceroni JA (2009) E-Work based collaborative optimization approach for strategic logistic network design problem. Comput Ind Eng 57(1):3–13. https://doi.org/10.1016/j.cie.2008.12.007
43. Moghaddam M, Nof SY (2013) Dynamic tool sharing with best matching protocols for efficient assembly line balancing. IFAC Proc Volumes 46(7):426–431. https://doi.org/10.3182/20130522-3-BR-4036.00045
44. Moghaddam M, Nof SY (2013b) Best matching and task administration protocols for effective demand and capacity sharing. Proceedings of ICPR22
45. Moghaddam M, Nof SY (2014) Combined demand and capacity sharing with best matching decisions in enterprise collaboration. Int J Prod Econ 148:93–109. https://doi.org/10.1016/j.ijpe.2013.11.015
46. Moghaddam M, Nof SY (2015) Balanceable assembly lines with dynamic tool sharing and best matching decisions—a collaborative assembly framework. IIE Trans 47(12):1363–1378. https://doi.org/10.1080/0740817X.2015.1027456
47. Moghaddam M, Nof SY (2015) Best-matching with interdependent preferences—implications for capacitated cluster formation and evolution. Decis Support Syst 79:125–137. https://doi.org/10.1016/j.dss.2015.08.005

48. Moghaddam M, Nof SY (2015) Real-time administration of tool sharing and best matching to enhance assembly lines balanceability and flexibility. Mechatronics 31:147–157. https://doi.org/10.1016/j.mechatronics.2014.10.001
49. Moghaddam M, Nof SY (2015d) Interdependent preferences-definitions and impacts on team formation. Proceedings of IIE annual conference
50. Moghaddam M, Nof SY (2016) Parallelism of Pick-and-Place operations by multi-gripper robotic arms. Robot Comput Integrated Manuf 42:135–146. https://doi.org/10.1016/j.rcim.2016.06.004
51. Moghaddam M, Nof SY (2016) Real-time optimization and control mechanisms for collaborative demand and capacity sharing. Int J Prod Econ 171:495–506. https://doi.org/10.1016/j.ijpe.2015.07.038
52. Moghaddam M, Nof SY (2017a) Best matching theory & applications. Springer International Publishing
53. Moghaddam M, Nof SY (2017) The collaborative factory of the future. Int J Comput Integr Manuf 30(1):23–43
54. Moghaddam M, Nof SY (2018) Collaborative service-component integration in cloud manufacturing. Int J Prod Res 56(1–2):677–691. https://doi.org/10.1080/00207543.2017.1374574
55. Nguyen WPV, Nair AS, Nof SY (2019) Advancing cyber-physical systems resilience: the effects of evolving disruptions. Procedia Manuf ICPR-25 39:334–340. https://doi.org/10.1016/j.promfg.2020.01.365
56. Nguyen WPV, Nof SY (2018) Resilience Informatics for cyber-augmented Manufacturing Networks (CMN): centrality, flow and disruption. Studies Inf Control 27(4):377–384
57. Nguyen WPV, Nof SY (2019) Collaborative response to disruption propagation (CRDP) in cyber-physical systems and complex networks. Decis Support Syst 117:1–13. https://doi.org/10.1016/j.dss.2018.11.005
58. Nguyen WPV, Nof SY (2020) Strategic lines of collaboration in response to disruption propagation (CRDP) through cyber-physical systems. Int J Prod Econ 230:107865. https://doi.org/10.1016/j.ijpe.2020.107865
59. Nof SY (1994) Information and collaboration models of integration (S. Y. Nof, Ed.; 1st ed.)
60. Nof SY (2003) Design of effective e-Work: review of models, tools, and emerging challenges. Prod Planning Control 14(8):681–703. https://doi.org/10.1080/0953728031000164783 2
61. Nof SY (2013) Sustainability and resiliency in supply networks. Proceedings of Asia pacific industrial engineering and management systems conference (APIEMS14)
62. Nof SY (2017) Collaborative control theory and decision support systems. Comput Sci J Moldova 25(2):115–144
63. Nof SY (2017b) IE674 lecture note: introduction to cyber in production. Purdue University
64. Nof SY (2019) From integration to augmentation, from interaction to collaborative control—IE/MS frontiers for future work and factories. Proceedings of APIEMS-IFPR_APR conference
65. Nof SY, Ceroni J, Jeong W, Moghaddam M (2015) Revolutionizing collaboration through e-work, e-Business, and e-Service (1st ed.)
66. Nof SY, Chen J (2003) Assembly and disassembly: an overview and framework for cooperation requirement planning with conflict resolution. J Intell Rob Syst 37(3):307–320. https://doi.org/10.1023/A:1025466401869
67. Nof SY, Weill R (1992) Collaborative coordination control (CCC) of distributed multimachine manufacturing. CIRP Ann 41(1):441–445. https://doi.org/10.1016/S0007-8506(07)61240-2
68. Peralta J, Anussornnitisarn P, Nof SY (2003) Analysis of a time-out protocol and its applications in a single server environment. Int J Comput Integr Manuf 16(1):1–13. https://doi.org/10.1080/713804980
69. Rad C-R, Hancu O, Takacs I-A, Olteanu G (2015) Smart monitoring of potato crop: a cyber-physical system architecture model in the field of precision agriculture. Agric Agric Sci Procedia 6:73–79. https://doi.org/10.1016/j.aaspro.2015.08.041

70. Rajan VN, Nof SY (1996) Cooperation Requirements Planning (CRP) for multiprocessors: optimal assignment and execution planning. J Intell Rob Syst 15(4):419–435. https://doi.org/10.1007/BF00437605
71. Rajan VN, Nof SY (1996) Minimal precedence constraints for integrated assembly and execution planning. IEEE Trans Robot Autom 12(2):175–186. https://doi.org/10.1109/70.488939
72. Reyes Levalle R, Nof SY (2015) A resilience by teaming framework for collaborative supply networks. Comput Ind Eng 90:67–85. https://doi.org/10.1016/j.cie.2015.08.017
73. Reyes Levalle R, Nof SY (2015) Resilience by teaming in supply network formation and re-configuration. Int J Prod Econ 160:80–93. https://doi.org/10.1016/j.ijpe.2014.09.036
74. Reyes Levalle R, Scavarda M, Nof SY (2012) Collaborative production line control for collaborative supply networks. Proc IFAC Symp Inf Control Problems Manuf 45(6):487–492. https://doi.org/10.3182/20120523-3-RO-2023.00316
75. Reyes Levalle R, Scavarda M, Nof SY (2013) Collaborative production line control: minimisation of throughput variability and WIP. Int J Prod Res 51(23–24):7289–7307. https://doi.org/10.1080/00207543.2013.778435
76. Reyes Levalle R, Scavarda M, Nof SY (2013b) Collaborative management of supply network operations. Proceedings of IIE annual conference
77. Reyes Levalle R, Scavarda M, Seok H, Nof SY (2012) Protocols for sustainable collaboration in supply networks. Proceedings of international conference on modeling, optimization & simulation
78. Scavarda M, Medel R, Lee S, Nof SY (2012) A market-based approach to adaptive reconfigurable supply networks. Proceedings of IIE annual conference
79. Scavarda M, Nof SY (2011) Cooperation requirement planning in supply decisions networks—the physical dimension. Proceedings of IIE annual conference
80. Scavarda M, Reyes Levalle R, Lee S, Nof SY (2017) Collaborative e-work parallelism in supply decisions networks: the chemical dimension. J Intell Manuf 28(6):1337–1355. https://doi.org/10.1007/s10845-015-1054-4
81. Scavarda M, Seok H, Nof SY (2017) The constrained-collaboration algorithm for intelligent resource distribution in supply networks. Comput Ind Eng 113:803–818. https://doi.org/10.1016/j.cie.2017.05.015
82. Scavarda M, Seok H, Puranik AS, Nof SY (2015) Adaptive direct/indirect delivery decision protocol by collaborative negotiation among manufacturers, distributors, and retailers. Int J Prod Econ 167:232–245. https://doi.org/10.1016/j.ijpe.2015.05.006
83. Seok H, Kim K, Nof SY (2016) Intelligent contingent multi-sourcing model for resilient supply networks. Expert Syst Appl 51:107–119. https://doi.org/10.1016/j.eswa.2015.12.026
84. Seok H, Nof SY (2011a) Decision support protocol for sustainability issues in supply networks. IIE Annual conference
85. Seok H, Nof SY (2011b) The HUB-CI initiative for cultural, education and training, and healthcare networks. Proceedings of ICPR21
86. Seok H, Nof SY (2012) Demand and capacity negotiated sharing by two manufacturers. Proceeding of IIE annual conference
87. Seok H, Nof SY (2014) Collaborative capacity sharing among manufacturers on the same supply network horizontal layer for sustainable and balanced returns. Int J Prod Res 52(6):1622–1643. https://doi.org/10.1080/00207543.2013.842016
88. Seok H, Nof SY (2014) Dynamic coalition reformation for adaptive demand and capacity sharing. Int J Prod Econ 147:136–146. https://doi.org/10.1016/j.ijpe.2013.09.003
89. Seok H, Nof SY (2018) Intelligent information sharing among manufacturers in supply networks: supplier selection case. J Intell Manuf 29(5):1097–1113. https://doi.org/10.1007/s10845-015-1159-9
90. Sreeram M, Nof SY (2021) Human-in-the-loop: role in cyber physical agricultural systems. Int J Comput Commun Control 16(2). https://doi.org/10.15837/ijccc.2021.2.4166
91. Suh SC, Tanik UJ, Carbone JN, Eroglu A (2014) Applied cyber-physical systems. Springer, New York, NY

92. Tkach I, Edan Y, Jevtić A, Nof SY (2013) Automatic multi-sensor task allocation using modified distributed bees algorithm. Proceedings of 2013 IEEE international conference on systems, man, and cybernetics, 1401–1406. https://doi.org/10.1109/SMC.2013.242
93. Tkach I, Edan Y, Nof SY (2011) A framework for automatic multi-agents collaboration in target recognition tasks. Proceedings of ICPR21
94. Tkach I, Edan Y, Nof SY (2012) Security of supply chains by automatic multi-agents collaboration. Proc IFAC Symp Inf Control Problems Manuf 45(6):475–480. https://doi.org/10.3182/20120523-3-RO-2023.00224
95. Tkach I, Edan Y, Nof SY (2013) Fault tolerant task allocation in a multi-sensor supply chain monitoring system. Proceedings of ICPR22
96. Tkach I, Edan Y, Nof SY (2017) Multi-sensor task allocation framework for supply networks security using task administration protocols. Int J Prod Res 55(18):5202–5224. https://doi.org/10.1080/00207543.2017.1286047
97. Tkach I, Jevtić A, Nof S, Edan Y (2018) A modified distributed bees algorithm for multi-sensor task allocation. Sensors 18(3):759. https://doi.org/10.3390/s18030759
98. US National Science Foundation (2006) Cyber-physical systems (CPS). https://www.Nsf.Gov/Pubs/2010/Nsf10515/Nsf10515.Htm#toc
99. Velásquez JD, Lara Gracia MA, Nof SY (2008) Systematic resolution of conflict situations in collaborative facility design. Int J Prod Econ 116(1):139–153. https://doi.org/10.1016/j.ijpe.2008.08.014
100. Velásquez JD, Nof SY (2008) A best-matching protocol for collaborative e-work and e-manufacturing. Int J Comput Integr Manuf 21(8):943–956. https://doi.org/10.1080/09511920802014896
101. Velásquez JD, Nof SY (2008) Integration of machine-vision inspection information for best-matching of distributed components and suppliers. Comput Ind 59(1):69–81. https://doi.org/10.1016/j.compind.2007.06.007
102. Velásquez JD, Nof SY (2009) Best-matching protocols for assembly in e-work networks. Int J Prod Econ 122(1):508–516. https://doi.org/10.1016/j.ijpe.2009.06.018
103. Williams NP, Liu Y, Nof SY (2002) TestLAN approach and protocols for the integration of distributed assembly and test networks. Int J Prod Res 40(17):4505–4522. https://doi.org/10.1080/00207540210155873
104. Williams NP, Liu Y, Nof SY (2003) Analysis of workflow protocol adaptability in TestLAN production systems. IIE Trans 35(10):965–972. https://doi.org/10.1080/07408170309342348
105. Yoon SW, Nof SY (2009) Distributed decision protocols for optimal demand and capacity sharing in collaborative enterprise networks. Proceeding of IIE annual conference
106. Yoon SW, Nof SY (2010) Demand and capacity sharing decisions and protocols in a collaborative network of enterprises. Decis Support Syst 49(4):442–450. https://doi.org/10.1016/j.dss.2010.05.005
107. Yoon SW, Nof SY (2011) Affiliation/dissociation decision models in demand and capacity sharing collaborative network. Int J Prod Econ 130(2):135–143. https://doi.org/10.1016/j.ijpe.2010.10.002
108. Yoon SW, Nof SY (2011) Cooperative production switchover coordination for the real-time order acceptance decision. Int J Prod Res 49(6):1813–1826. https://doi.org/10.1080/00207540903567325
109. Zhang L, Zhong H, Nof SY (2015) Adaptive fuzzy collaborative task assignment for heterogeneous multirobot systems. Int J Intell Syst 30(6):731–762. https://doi.org/10.1002/int.21725
110. Zhong H, Nof SY (2013) Collaborative design for assembly: the HUB-CI model. Proceedings of ICPR22
111. Zhong H, Nof SY (2014) DLOC complex network model for supply network disruption response. Proceedings of ICPR-AEM
112. Zhong H, Nof SY (2015) The dynamic lines of collaboration model: collaborative disruption response in cyber–physical systems. Comput Ind Eng 87:370–382. https://doi.org/10.1016/j.cie.2015.05.019

113. Zhong H, Nof SY (2020) Dynamic lines of collaboration disruption handling & control (S. Y. Nof, Ed.; 1st ed.)
114. Zhong H, Nof SY, Berman S (2015) Asynchronous cooperation requirement planning with reconfigurable end-effectors. Robot Comput Integr Manuf 34:95–104. https://doi.org/10.1016/j.rcim.2014.11.004
115. Zhong H, Nof SY, Ozsoy E (2015) Co-Viz: matching tools in collaborative visual analytics. Proceedings of IIE annual conference
116. Zhong H, Ozsoy E, Nof SY (2016) Co-Insights framework for collaborative decision support and tacit knowledge transfer. Expert Syst Appl 45:85–96. https://doi.org/10.1016/j.eswa.2015.09.036
117. Zhong H, Reyes Levalle R, Moghaddam M, Nof SY (2015) Collaborative intelligence—definition and measured impacts on internetworked e-work. Manag Prod Eng Rev 6(1):67–78. https://doi.org/10.1515/mper-2015-0009
118. Zhong H, Wachs JP, Nof SY (2013) HUB-CI model for collaborative telerobotics in manufacturing. Proc IFAC Intell Manuf Syst 46(7):63–68. https://doi.org/10.3182/20130522-3-BR-4036.00059

Chapter 2
Cyber-Collaborative Protocols in Precision Agriculture and Agricultural Robotic Systems

Abstract This chapter presents the environment setting of the study of CCP: The agricultural environment is used as the domain example here and also later in the book. First, we present our motivation and reasons why we select agriculture, specifically Precision Agriculture (PA). Then, an Agricultural Robotic System (ARS) design model is presented. After that, we explain four main assumptions, used in the study, to provide a general, common setting, and explain the benefits of utilizing CCP for PA. Lastly, limitations of CCP for improving PA are discussed.

2.1 Motivation of Precision Agriculture (PA)

Farmers nowadays grow crop plants in a greenhouse to ensure food production and eliminate unwanted, disrupting factors. Crop plants are treated in near-perfect control conditions to grow and maximize their yield [7]. Many of the farmers, however, are far from applying precision-control, as fluctuation in environmental conditions such as humidity, temperature, airflow, sunlight, etc., causes stress in crop plants [13]. In many cases, a crop plants' stress may lead to diseases and eventually results in yield loss. Between 1980 and 2012, crop plants' stresses and diseases generated by environmental fluctuation caused damages evaluated at more than $200 billion worth in the U.S. agricultural industry [26].

Any abnormal status of crop plants (e.g., stress) must be detected, localized, and treated early to minimize the outcome of yield loss. Traditionally, farmers assume crop plants in the growing fields have homogeneous properties [22], therefore they evenly apply farming resources such as water, fertilizer, and pesticide to entire agricultural areas to treat crop plants and reduce the loss [17]. This practice wastes farming resources, money, and time. Research also indicates that improper watering of crop plants may fail to boost crop plants yield [5]. Worse yet, over- or under-fertilizing crop plants will damage financial return and crop plants production [23]. Hence, uniform resource applications may not be the optimal practice for the future of agriculture [3].

P. O. Dusadeerungsikul and S. Y. Nof, *Cyber-Collaborative Algorithms and Protocols*, Automation, Collaboration, & E-Services 15,
https://doi.org/10.1007/978-3-031-56196-2_2

Currently, there is an alternative to traditional agricultural management, called Precision Agriculture (PA). PA aims to apply only the right amount of resources at the right time and at the right location [18], driven by information and communication technologies [25]. With such objective, PA can ensure the proper treatment for each crop plants while minimizing production cost. In addition, PA requires a data acquisition system including sensors, algorithms, and tools for analysis to adjust to dynamic changes in unstructured environment [2]. Thereby, the resources, money, and time in PA can be utilized precisely and effectively [27]. Specifically, one of the key elements required for effective PA is the ability to early detect stress and disease, so that the cost of treatment can be reduced significantly [16].

As we can see, PA has multiple inputs, outputs, and interactions. To deal with such complexity, smart agriculture and advanced technologies for PA, called Agriculture Robotic System (ARS), are crucial to enhance and achieve the advantages. ARS does not only precisely operate framing activities but also enables farmers to work and monitor crop plants remotely, with the support of Cyber-Physical System (CPS), including intelligent control. Its task is to enable the new concept of future farming [14]. Therefore, PA with ARS can reduce overall operating costs and maintain crop plants production objectives, thus ensuring human food security.

2.2 Agricultural Robotic System (ARS)

Agricultural Robotics System (ARS) refers to a smart agricultural system utilizing the strength of multi-agents to optimize system performance with limited resources. The definition of ARS has addressed both objectives and challenges. ARS requires advanced technologies such as agricultural robotics, multi-spectrum sensors, etc. Also, the emerging solutions such as Cyber- Collaborative Protocol (CCP), Cyber-Physical System (CPS), Internet of Things/Internet of Services (IoT/IoS), Collaborative Control Theory (CCT), and Precision Collaboration (PCol) must be integrated and utilized. The CCP is the primary focus in this book. The other topics in this list are explained too as applicable in this book, and in other referenced publications.

ARS consists of human operators, a robot arm mounted on an autonomous mobile cart, multiple sensors, algorithms, knowledge-based information, and other supporting tools for intelligent control [8]. Since each agent and algorithm have their defined roles, strengths, and weaknesses, CCPs are the critical tools to synchronize and harmonize all of the agents and others to optimally perform together their tasks. For example, sensors have the advantages of quickly collecting essential data, which will be analyzed for a suitable treatment for each location [12]. Sensors, however, cannot work by themselves and need agricultural robotics and human operators to complete the operation [15]. Besides, an algorithm that provides decision output for a particular scenario needs input from relevant sensors. Therefore, CCP is an essential tool, providing linkage mechanism managing agents, algorithms, and knowledge-based information. It can be viewed as an AI operations manager. Several CCPs can interact to generate integrated optimal, orchestrated operations of the entire ARS.

This book aims to analyze and design CCP for each level of ARS, namely, system level, agent level, algorithm level, auxiliary level, and system-of-systems level. System analytics and operation research have been utilized for more than five decades to help the decision-maker make a better decision. We choose an agricultural system as the domain of application since it is, by nature, relatively unstructured and prone to system errors [1]. Hence, it can represent well multiple levels of challenge in system management and protocol design. Issues and challenges in the agricultural system domain are addressed to help test the validity, applicability and effectiveness of the optimal operations in an unstructured and uncertain system of significant value to humankind.

2.2.1 *Assumptions About the Agricultural Systems*

We postulate four general assumptions that are used throughout the studies. In this section, each assumption is explained in detail.

2.2.2 *Assumption 1: System Scope Assumption*

The first assumption deals with the system scope. There are three main types of agents in ARS: humans, robots, and sensors. Humans aided by the expert system are the decision-makers ultimately responsible for solving unexpected and complex real-time problems. An autonomous mobile robot cart is guided to assigned locations for inspecting crop plants. The autonomous cart is mounted with a robot arm (manipulator) attached with multiple sensors that serve as detecting agents. Over the agricultural field, crop plants are selected for inspection (and subsequently for any required and designated treatment) according to a sampling plan. Random sampling in the simplest sampling strategy.

The greenhouse layout, which is the research environment in this study, is illustrated in Fig. 2.1. Human–robot base point is the location where a mobile robot starts, and human experts are located. When a robot with sensors travels into the greenhouse, data is collected and transmitted to the base point computers for further analysis.

2.2.3 *Assumption 2: Parameter Assumption*

The second assumption relates to the input parameters of the model. Input parameters such as the number of agents, agent's movement speed and capabilities, greenhouse structure, and other modeling features are assumed to be known. The reason they

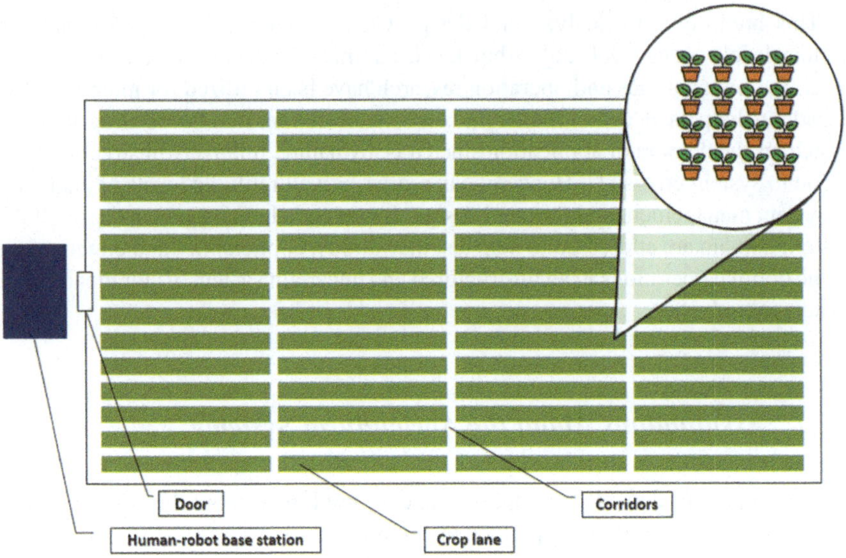

Fig. 2.1 Greenhouse layout

are assumed to be known is because the mentioned parameters can be measured and calculated ahead of the system operations.

For simplicity, all the parameters listed above are assumed to be constant and not changing over time. There is no learning effects impact on the parameters for machine and human involvement in the system. In other words, with this assumption, the system analyses are for the worst-case scenario, as it is assumed that agents' performance will not be improved overtime (even though we know it does improve).

On the other hand, the uncertainty of input parameters is considered in the model. In particular, the distribution of input parameters represents such uncertainty. Parameters such as mobile cart travel speed, sensor inspection time, and robot arm movement speed are assumed to be normally distributed.

Parameters such as Conflicts and Errors (C&E), location of the stressed crop plants are assumed to exist in the system, originally with unknown location and value. The assumption reflects the real scenario because the parameters have existed in the agricultural system, but the system designers may not know their location and value.

2.2.4 *Assumption 3: Agent Assumption*

Assumption 3 is the agent assumption. The assumption describes the agent's capabilities and role of agent. In this book, there are three main agents (human, robot, and sensors). Each agent has their role and responsibility for performing a specific

Table 2.1 Agents and their roles

Agent	Role of agent
Human with the expert system	Analyze data and decide as assigned; Decision-maker who solves unexpected real-time situations;
Autonomous cart with Robot arm	Travel into the greenhouse; Approach crop plant's parts; Compute and respond to given commands;
Sensors	Collect data (non-contact); Transmit collected data

task. A human with an expert system is the decision-maker who is assumed to be superior to other agents in terms of real-time decision-making [24] and is responsible for deciding the status of the crop plants, solving unexpected real-time situations, and commanding the robot as needed. An autonomous mobile cart, mounted with a robot arm, that are together good at executing repetitive tasks [19]. The cart (with the mounted robot arm) will move into the greenhouse to visit assigned locations. Then, the robot arm will approach part of the crop plants to perform designated inspection by sensors. Sensors which can identify stress or primary disease characters [10, 11] will collect data such as photos and videos of the crop plant's part for further analysis [21, 25]. After that, the sensors will transmit the information to the expert system for a necessary decision, if needed. Table 2.1 summarizes the main agents and their roles in the ARS.

Based on the task descriptions mentioned earlier, each agent is assumed to work on its own. For example, a mobile robot with an autonomous cart can work by its capability. It can move and approach the designated crop plant's part regardless of the readiness of sensors. Also, agents such as sensors and IoT/IoS may require support from other agents to complete the inspection process, but the sequence condition between agents' operations does not exist in the system a priori. (It is calculated and determined by the collaborative control system, as explained later.)

With these characteristics, the agents can work in parallel, supporting the CCT e-Work Parallelism principle [20]. This work parallelism can be significantly beneficial, but may cause inferior performance unless careful CCPs are designed as part of the collaborative control system. Furthermore, because of the parallelism, there might be a higher rate of C&E in the system [4]. Hence, CCP must take such concerns into account during the protocol design phase, as explained in this book.

2.2.5 Assumption 4: Terminology Assumption

The last assumption is the terminology assumption. The critical terms are assumed as follows.

1. *Agricultural Robotic System, ARS*, refers to the entire monitoring system, not just a particular agent. The reason is that a particular agent cannot work to deliver the desired outcome by itself. The system agents need to work collaboratively

to achieve the system's goal. Hence, ARS must refer to the entire monitoring system as a team.

2. *Humans* or *Human agents* refer to farmers, researchers, experts with an expert system and information-based. Human agents usually have higher knowledge than other agents so that humans can analyze and evaluate most unexpected, often rare situations during the system operation [24].
3. *Robot* refers to the autonomous robot cart, robot's arm, and other manipulators that move into the greenhouse and approach crop plant's parts in multiple directions. Also, robots need to connect and respond to the human command in emergency cases. The term robot also refers to the treatment robots dispatched for the precise treatments needed and administered in case of detected stresses or diseases.
4. *Sensors* refer to multi-spectral cameras, thermal sensors, video cameras, high-definition cameras, and other IoT/IoS devices. Sensors are responsible for collecting data such as images, video, temperature, and humidity and sending their gathered information for further analysis.

2.3 In Which Aspects Can CCP Significantly Improve the Agricultural System?

Because food security is an increasingly important issue recently [9], preventing production yield loss of crop plants is critical. Crop plants usually develop stress in the initial state of diseases, usually resulting in yield loss [6]. Detecting the stress as early as possible before diseases emerge, and treating the crop plants precisely with thoughtfully applied resources is the ultimate goal of ARS. At the same time, it is also a critical challenge. Hence, to achieve the objective of ARS, CCP is utilized to improve the system from the following perspectives:

1. *Collaboration perspective*: The CCP (with the integration of CPS) can provide the overall agricultural CPS framework combining agents, algorithms, and knowledge-based information to work effectively and harmoniously, and facilitate real-time communication and control for the ARS. (Chapter 3 demonstrates this perspective.)
2. *System objective perspective*: CCP can connect agents, algorithms, and knowledge-based information to maximize system objectives under the CPS framework. The examples of system objectives include responding based on real-time information (e.g., current, precise health characteristics of each individual crop plant), effectively utilizing available resources (e.g., time, agents, knowledge) for the earliest detection of stress in crop plants. (Examples are given in Chaps. 4, 5, and 6)
3. *Multiple team coordination perspective*: CCP can be applied to manage multiple teams in ARS. For example, apart from the monitoring team, after detecting the stress, ARS can deploy a treatment team to properly recover the unhealthy crop plant. (Chapter 7 provides a detailed example.)

We can see that CCP can be applied to multiple layers of ARS to improve the agricultural system and its productivity. The following chapters will present both a theoretical framework and practical implications in various settings.

2.4 Limitations of CCP for Agricultural Systems

Even though CCP can significantly improve the ARS capabilities and yield in several perspectives, it is not a one-size-fit-all tool. Like other methodologies, there are also limitations that we need to take into account.

1. *Changes in environment*: Because ARS operates in agricultural systems (even if it is a well-designed greenhouse structure), the environment is subject to changes overtime. Crop plant's leaves or stems, for example, can change when airflow or sunlight shifts their direction. Such limitations require engineers and researchers to design more complex CCP to address the changes.
2. *Changes in system components*: The CCP is designed base on assumption discussed above in Sect. 2.3. In practice, however, some of them may require a relaxation. The number of agents, for example, is designed and fixed in advance. If a greenhouse is relatively larger than the original design, a single robot may not cover all areas. In such scenario, we may consider using multiple robots to fulfill the monitoring and treatment tasks. This challenge requires researchers to review and revise the current CCP for an effective operation at a larger scale.
3. *Changes in input*: In agricultural settings, inputs such as knowledge and expertise about crop plants' diseases and stresses, might vary within a short period. Such deviation can cause a non-optimal solution derived from CCP. In fact, diseases, even when captured by sensors may quickly erupt and become more severe before the treatment team can arrive. To minimize and address this limitation, researchers and engineers need to utilize the advanced technologies that enable real-time communication and control.

Most often, for essential crop plants, knowledge, expertise, pathological evidence and precision treatments procedures are well researched and developed. Additionally, sophisticated CCPs can be designed with machine learning, adaptive and predictive functions to overcome some of the limitations.

References

1. Bechar A, Edan Y (2003) Human-robot collaboration for improved target recognition of agricultural robots. Indus Robot: Int J 30(5):432–436. https://doi.org/10.1108/01439910310492194
2. Bloch V, Bechar A, Degani A (2017) Development of an environment characterization methodology for optimal design of an agricultural robot. Indus Robot: Int J 44(1):94–103. https://doi.org/10.1108/IR-03-2016-0113

3. Bock CH, Poole GH, Parker PE, Gottwald TR (2010) Plant disease severity estimated visually, by digital photography and image analysis, and by hyperspectral imaging. Crit Rev Plant Sci 29(2):59–107. https://doi.org/10.1080/07352681003617285
4. Chen XW, Nof SY (2012) Conflict and error prevention and detection in complex networks. Automatica 48(5):770–778. https://doi.org/10.1016/j.automatica.2012.02.030
5. Drechsel P, Heffer P, Magen H, Mikkelsen R, Singh H, Wichelns D (2015) Managing water and nutrients to ensure global food security, while sustaining ecosystem services
6. Dusadeerungsikul PO, Liakos V, Morari F, Nof SY, Bechar A (2020) Smart action. In: Agricultural internet of things and decision support for precision smart farming, pp 225–277. https://doi.org/10.1016/B978-0-12-818373-1.00005-6
7. Dusadeerungsikul PO, Nof SY (2019) A collaborative control protocol for agricultural robot routing with online adaptation. Comput Indus Eng 135. https://doi.org/10.1016/j.cie.2019.06.037
8. Dusadeerungsikul PO, Nof SY (2021) A cyber collaborative protocol for real-time communication and control in human-robot-sensor work. Int J Comput Commun Control 16(3):1–11. https://doi.org/10.15837/ijccc.2021.3.4233
9. Food and Agriculture Organization of the United Nations (2019) World food and agriculture—statistical pocketbook 2019. FAO. https://doi.org/10.4060/CA6463EN
10. Franke J, Gebhardt S, Menz G, Helfrich H-P (2009) Geostatistical analysis of the spatiotemporal dynamics of powdery mildew and leaf rust in wheat. Phytopathology 99(8):974–984. https://doi.org/10.1094/PHYTO-99-8-0974
11. Franke J, Menz G (2007) Multi-temporal wheat disease detection by multi-spectral remote sensing. Precis Agric 8(3):161–172. https://doi.org/10.1007/s11119-007-9036-y
12. Gongal A, Amatya S, Karkee M, Zhang Q, Lewis K (2015) Sensors and systems for fruit detection and localization: a review. Comput Electron Agric 116:8–19. https://doi.org/10.1016/j.compag.2015.05.021
13. Guo P, Dusadeerungsikul PO, Nof SY (2018) Agricultural cyber physical system collaboration for greenhouse stress management. Comput Electron Agric 150:439–454. https://doi.org/10.1016/j.compag.2018.05.022
14. Herlitzius T (2017) Automation and robotics—the trend towards cyber physical systems in agriculture business. https://doi.org/10.4271/2017-01-1932
15. Ko MH, Ryuh B-S, Kim KC, Suprem A, Mahalik NP (2015) Autonomous greenhouse mobile robot driving strategies from system integration perspective: review and application. IEEE/ASME Trans Mechatron 20(4):1705–1716. https://doi.org/10.1109/TMECH.2014.2350433
16. Mahlein A-K (2016) Plant disease detection by imaging sensors—parallels and specific demands for precision agriculture and plant phenotyping. Plant Dis 100(2):241–251. https://doi.org/10.1094/PDIS-03-15-0340-FE
17. Mandal D, Ghosh SK (2000) Precision farming—the emerging concept of agriculture for today and tomorrow. Curr Sci 79(12):1644–1647
18. Mulla DJ (2013) Twenty five years of remote sensing in precision agriculture: key advances and remaining knowledge gaps. Biosys Eng 114(4):358–371. https://doi.org/10.1016/j.biosystemseng.2012.08.009
19. Nair AS, Bechar A, Tao Y, Nof SY (2019) The HUB-CI model for telerobotics in greenhouse monitoring. Procedia Manuf ICPR 25:414–421. https://doi.org/10.1016/j.promfg.2020.01.385
20. Nof SY, Ceroni J, Jeong W, Moghaddam M (2015) Revolutionizing collaboration through e-Work, e-Business, and e-Service, 1st edn
21. Nutter FW Jr, Littrell RH, Brenneman TB (1990) Utilization of a multispectral radiometer to evaluate fungicide efficacy to control late leaf spot in peanut. Phytopathology 80(1):102–108
22. Oerke E-C, Fröhling P, Steiner U (2011) Thermographic assessment of scab disease on apple leaves. Precision Agric 12(5):699–715. https://doi.org/10.1007/s11119-010-9212-3
23. Ribaudo M, Hansen L, Livingston MJ, Mosheim R, Williamson J, Delgado J (2011) Nitrogen in agricultural systems: implications for conservation policy. SSRN Electron J. https://doi.org/10.2139/ssrn.2115532

24. Sreeram M, Nof SY (2021) Human-in-the-loop: role in cyber physical agricultural systems. Int J Comput Commun Control 16(2). https://doi.org/10.15837/ijccc.2021.2.4166
25. Stafford JV (2000) Implementing precision agriculture in the 21st century. J Agric Eng Res 76(3):267–275. https://doi.org/10.1006/jaer.2000.0577
26. Suzuki N, Rivero RM, Shulaev V, Blumwald E, Mittler R (2014) Abiotic and biotic stress combinations. New Phytol 203(1):32–43. https://doi.org/10.1111/nph.12797
27. Thompson NM, Bir C, Widmar DA, Mintert JR (2019) Farmer perceptions of precision agriculture technology benefits. J Agric Appl Econ 51(1):142–163. https://doi.org/10.1017/aae.2018.27

Chapter 3
Cyber-Collaborative Protocols for Cyber-Physical Systems (CCP-CPS)

Abstract In this chapter, we present the application of Collaborative Control Protocol (CCP) with a high-level view, an overview of a monitoring system. The CCP is applied to improve the Cyber-Physical Systems (CPS)-oriented framework and workflow for agricultural greenhouse stress management, at the system level of Agricultural Robotic System (ARS). We develop the Collaborative Control Protocol for Cyber-Physical Systems (CCP-CPS), aiming to enable collaboration among agents in ARS for regular as well as unexpected situations. Humans, robot, and sensors are the three key agents in the ARS, and the protocol is the foundation of the task collaboration and connection among agents collaborating in the system. In this chapter, we discuss the design of the protocol and its unique features. Then, the protocol is validated using experiments to investigate the effects of CCP in CPS, comparing with an alternative design, the current procedure (practice). Note that the purpose of the protocols and algorithms built in the following chapter is to enable and support the enhancement of the CCP-CPS performance and accomplishment of its roles.

3.1 Design of Cyber-Collaborative Protocol for Cyber-Physical System (CCP-CPS)

A CPS-oriented framework and workflow are modified from the Monitoring, Detecting, and Responding—Cyber-Physical System framework (MDR-CPS) developed by Guo et al. [12]. It has been improved by a group of researchers (see Dusadeerungsikul et al. [7]) because an Agricultural Robotic System (ARS) requires the CPS to collaboratively operate, effectively communicate, and share information in real-time with all involved participating agents. The primary goal of this chapter is to illustrate the importance of Cyber-Collaborative Protocol (CCP) with Cyber-Physical System (CPS) integration in a greenhouse.

P. O. Dusadeerungsikul and S. Y. Nof, *Cyber-Collaborative Algorithms and Protocols*, Automation, Collaboration, & E-Services 15,
https://doi.org/10.1007/978-3-031-56196-2_3

3.1.1 CPS Framework for ARS

To effectively monitor crop plants in greenhouses, CPS is necessary. The CPS framework is depicted in Fig. 3.1. In addition, Fig. 3.2 shows the deployment of the CPS framework for ARS, operating in a greenhouse at Israel's Volcani Institute. This system includes a mobile robotic cart, developed at Volcani's Agricultural Robotics Lab, which is equipped with a robotic arm and a variety of monitoring sensors. Because of CPS, this robot can be remotely controlled from the Purdue PRISM Center. In addition, the ARS incorporates sensor data analysis and machine learning algorithms, a product of the University of Maryland's Bio-Imaging and Machine Vision Lab. The ARS also integrates online local resources, remote farmers, and knowledge-based information developed by plant biologists.

Figure 3.3 which is derived from the CCP design, the system has two collaborative automation and cyber-control layers: Task layer (T layer) and Resource layer (R layer). At time t, the two layers are connected and communicate with the physical elements of the CPS. R layer includes the agent or resource n which needs to perform a given task m in the T layer. Since agents in R layer are distributed and decentralized, they need efficient communication and shared information in order to complete each of the assigned tasks.

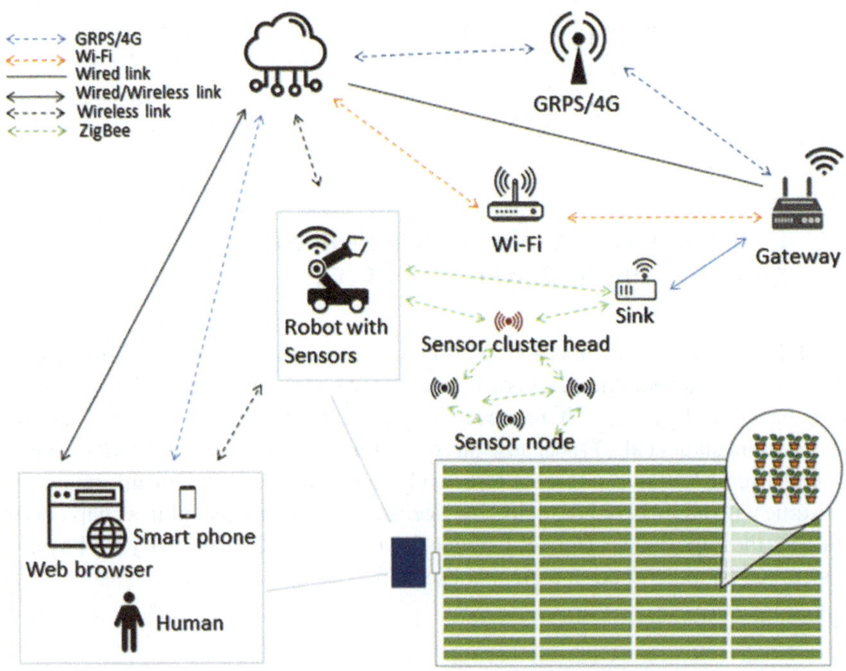

Fig. 3.1 CPS framework for monitoring crop plant conditions in a greenhouse

Fig. 3.2 The implementation of CPS framework for ARS at Volcani Institute, Israel [13]

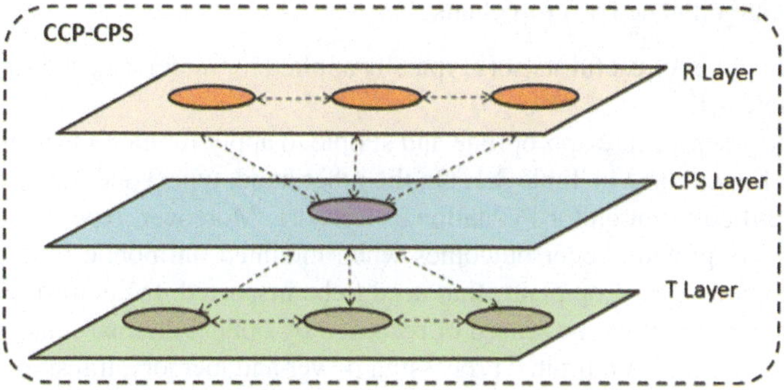

Fig. 3.3 Framework for greenhouse monitoring system

The essential communications between agents and tasks are depicted by arrows (Fig. 3.3). Based on these communications, the CPS middle layer (often called Middleware) can create a stable and capable connection between the *T* and *R* layers. Wireless Sensor Network (WSN), Cloud platform, Transmission mode, agricultural robots, and other actuators are the four key components of the CPS network of collaboration. The following sections explain each part of the CPS for agricultural monitoring tasks in greater detail.

3.1.1.1 Wireless Sensor Network (WSN) Deployment

Sensors are vital parts of the overall environmental monitoring and control system [18] and in agricultural automation [23]. The use of wireless sensor technologies is relatively more feasible in a greenhouse than in outdoor applications since crop plant conditions within greenhouses are relatively more regulated [27]. The WSN deployment refers to the installation of sensor nodes in greenhouses to provide information about environmental parameters that influence crop plant production. In addition, the WSN receives and transmits signals from the robot's local sensors.

In the model, there are various types of heterogeneous agricultural sensors that track environmental parameters or abiotic stresses such as humidity, temperature, pressure, CO_2 density, sunlight density, and water levels. The WSN has a role in the three key monitoring functions:

(1) Monitoring diseases or static biotic stresses of crop plants, for instance, bacteria, fungi, and viruses;
(2) Monitoring moving biotic stresses of crop plants, such as insects, rats, and other unexpected intruders;
(3) Monitoring chemical effects, such as crop plants nutrients, soil PH value, and pesticide influence on crop plants.

Different types of useful sensors typically applied for monitoring are summarized next in Table 3.1.

Stationary sensors are appropriate and simple to apply for monitoring the parameters type 1 and type 2 in Table 3.1. On the other hand, type 3 and 4 parameters are typically difficult to monitor by stationary sensors. Moreover, type 3 and 4 monitoring sensors provide better outcomes when mounted on mobile robots, and the robots navigate to the crop plants that need to be inspected. In contrast, parameters of type 5 are normally to be sensed or recorded by optical/thermal sensors or high-definition cameras. With limited processing power and memory, transmitted images from stationary sensors may not be clear enough for reliable analysis. Therefore,

Table 3.1 Sensor types for monitoring different parameters

Type	Monitoring categories	Example parameters	Type of sensor
1	Environmental parameters	Temperature, humidity, CO_2, Sunlight density, Water level, Airflow	Stationary sensor
2	Soil and ground parameters	Soil humidity, Pesticide level, Soil temperature, Soil PH value, Soil compaction	
3	Growth parameters	Stems and leaves parameters, Humidity, Fruit swelling, Angle of growth	Sensor mounted on mobile robot
4	Healthiness parameters	Crop plants stress, Bacteria, Fungi, Viruses	
5	Insects and other intruders parameters	Insects, Rats, Bugs, Worms	Stationary sensor and HD camera

sensor type 5 is suitable for either a cyber-control system which triggers humans, or humans that dispatch a robot to collect additional information with sensors mounted on a mobile robot. In certain applications, a drone may serve as a mobile robot.

3.1.1.2 Cloud Platform

To simplify the scenario, our agricultural cloud platform is centered on a few clusters. It consists of two components: cloud storage and cloud computing/expert systems. These components do not only store a large amount of sensor-acquired data, but also offer services such as crop plant disease analysis and diagnosis, intruder detection, and stress identification. The agricultural cloud is built on cloud computing theory and technology [12, 25]. The advantages of the system are (1) low cost of retaining, sharing, and sustaining resources, (2) reducing farmer burden; and (3) minimizing human errors. Using a trained database, an expert system will assess the new information received from sensors, to determine whether there are stresses in crop plants. If there is a disease on crop plant leaves, cameras will capture and upload images to the cloud platform, which is then processed by an expert system, and an appropriate notification is sent to farmers. This is the state-of-the-art in Machine Learning and Artificial Intelligence (ML&AI) techniques, working with expert systems. ML&AI research has been addressed in several PA topics [1], for example, crop plant image classification and recognition [15, 21], smart irrigation [11], data mining and knowledge extraction [6, 19], prediction and detection of biotic stress in crop plants [4, 24], crop plant yield evaluation [9], and predicting environmental factors [17, 20].

The current chapter focuses on how CCP and collaborative workflow can be used for optimally managing the agricultural CPS to track, detect, and respond in time to agricultural stresses. The application of ML&AI in ARS and Precision Agriculture (PA) is presented later in Chap. 6.

3.1.1.3 Transmission Mode

The network layer performs routing and data aggregation. In the CPS framework (as seen in Fig. 3.1), sensors transmit data to sink nodes, cameras, and sink nodes connect to the gateway through wireless links. The gateway communicates with the agricultural cloud through GPRS/4G, Internet, Wi-Fi, or local area networks. Furthermore, human operators or farmers may use a web browser or smartphone to access agricultural data, allowing them to solve problems in real-time. Specific communication transmission technologies are shown in Table 3.2.

3.1.1.4 Agricultural Robots and Other Actuators

Sensors track the physical environment [3, 28], and actuators allow physical processes to take place [22]. The terminal computing module comprises the basic

Table 3.2 Communication transmission technologies applied in CCP-CPS

Communication methods	Applications and usage	Connection attribute
Zigbee	Between sensors, Sensor to sink node, robot to sensor, robot to sink node	Wireless
Wi-Fi, LAN, Internet, 4G, GPRS	Agricultural gateway to cloud computing, Cloud computing to web browser or smartphone, robot to agricultural gateway	Wire or wireless

execution rules of an actuator and has limited real-time data storage space. To illustrate, in case the temperature, humidity, and solar radiation do not match the preset parameters, the software-hardware interfaces will enable the corresponding hardware/actuation equipment to adjust automatically. This is the reason we pay close attention to agricultural robots and actuators in our design model. An agricultural robot is designed to assist detection in specific circumstances for specific stresses. Though sensors may perform most of the monitoring work and obtain images or photographs, they are limited by power, location, orientation, and transmitting capacity [5]. For example, suppose the agricultural robot with mounted sensors discovers an unusual condition, such as fungi on crop plant leaves, based on photos obtained by cameras. The common procedure is that sensors will transmit such information to humans to determine what type of stress or disease occurs and at which level of severity. In some cases, such initial information may not be sufficient for decision-making. Thus, farmers need a mobile robot to revisit the given location. Once arrived, the cameras mounted on the robot will take additional photos or record a video at several different angles and send them to an agricultural expert system for detailed analysis. That is a rational requirement for including a robot in the system since it can not only be equipped with sensors to increase computational power, memory capability, and electrical power, but also approach and image crop plant parts that humans require for better quality decision-making.

According to the descriptions above, we can determine the computational structure of an agricultural robot for CCP-CPS, as shown in Fig. 3.4. The robot computer's objective is to run the software required for interacting with the robot platform and sensors, sensor data processing, mission planning and execution, navigation, implementation control, user interface, and network communication.

We have discussed so far the main elements of CPS for PA. In the following section, we begin further explanation of the CCP-CPS design.

3.1.2 CCP-CPS Design

The workflow for the CCP-CPS framework is shown above in Fig. 3.1. The details of the workflow can be described as follows.

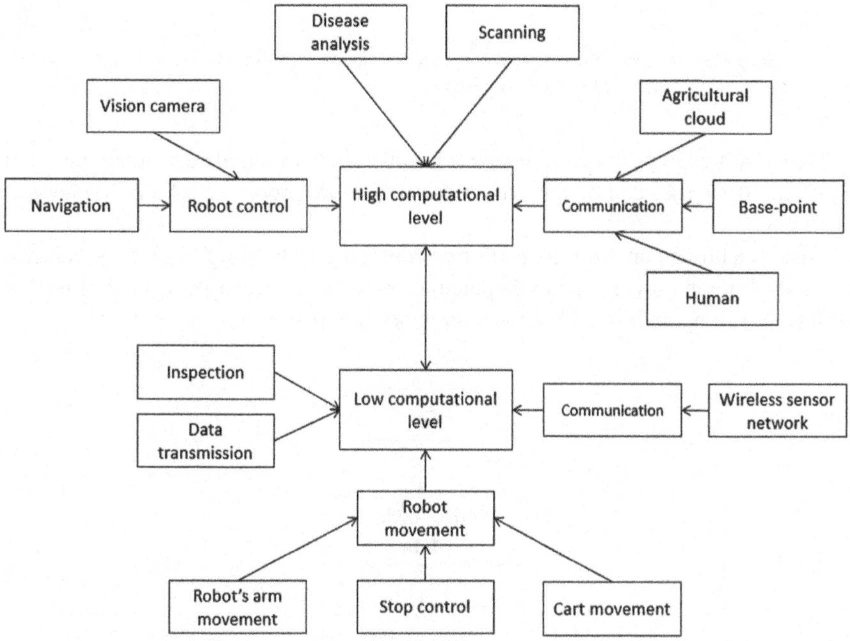

Fig. 3.4 Computational structure for CCP-CPS in ARS

CCP-CPS workflow		
Step 1	CCP-CPS system initialization	
Step 2	Sensors have been triggered to monitor the greenhouse	
Step 3	Sensed data have been transmitted to the agricultural cloud	
Step 4	Data have been saved in the agricultural cloud storage	
Step 5	An agricultural expert system has analyzed the collected data	
Step 6	If no stresses have been found, no action is needed	
Step 7	If confirmed stresses have been detected, the corresponding mechanism will be triggered	
Step 8	If an agricultural expert system cannot confirm because photos are unclear or insufficient, then produce an informed alarm signal and deliver it to humans	
	Step 8.1	Upon receiving the alarm signal, humans instruct a robot to navigate to a given area to collect more data
	Step 8.2	The robot is launched to navigate when it receives an order from humans

(continued)

(continued)

	Step 8.3	The robot takes photos or records videos regarding the human order when it reaches the given crop plants

Note that for Steps 4 and 5, the agricultural expert system detects again based on new data; if stresses can be confirmed, the procedure proceeds to Step 7, otherwise to Step 8.

Also, if a human operator receives two consecutive alarm signals for a given crop plant, the human may decide to dispatch a robot for further inspection. A flowchart with conversion conditions describes the workflow (Fig. 3.5).

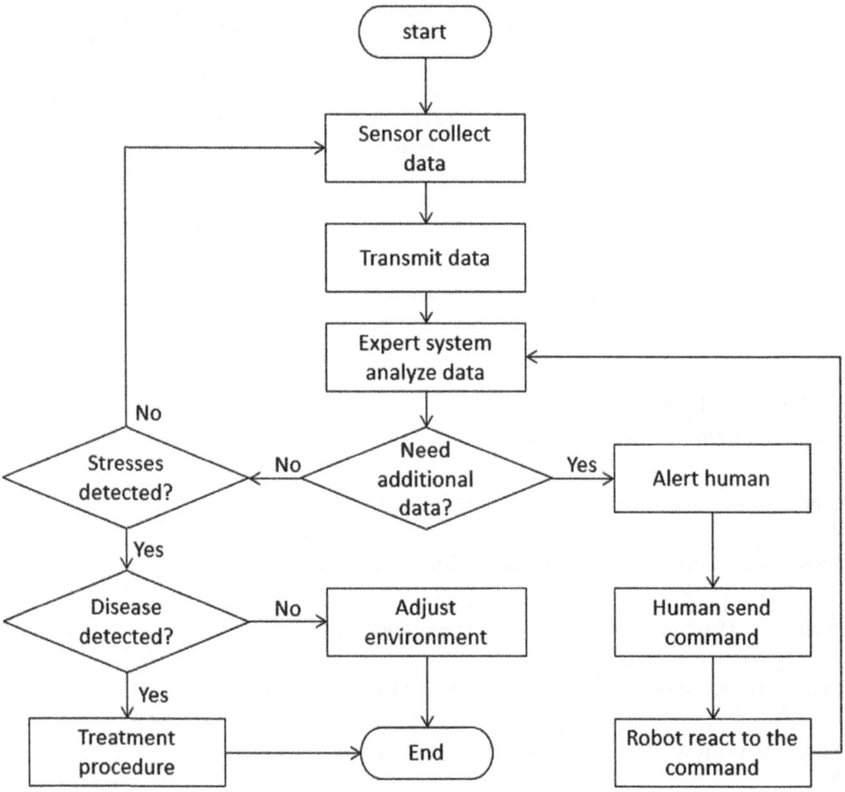

Fig. 3.5 CCP-CPS workflow

3.1.3 *Collaborative Control Theory in CCP-CPS*

In CCP-CPS, we applied Collaborative Control Theory (CCT) to design the collaborative operations of agents working together. We now move into the theory behind the design of this CCP-CPS protocol.

3.1.3.1 Collaborative Architecture for CCP-CPS

Collaboration is needed to successfully utilize the system's resources for timely accomplishment of required tasks. Collaboration can result in a more effective system with less errors [2, 16]. Formally, collaboration in CCP-CPS is classified into five levels in the model: collaborative sensing, collaborative data processing, collaborative communication, collaborative acting, and collaborative control (Fig. 3.6).

- **Collaborative sensing**: Multiple sensors collect by sensing environment data and current crop plant physical attributes data. Different sensors are responsible for different parameters or data [8].
- **Collaborative processing**: This entails a mobile robot equipped with sensors to monitor and inspect crop plants in suitable directions. When data needs to be gathered, a mobile robot equipped with sensors must reach the given target and approach the parts of the crop plant and environment, such as leaf, stem, and soil. This collaboration is crucial, especially when the operating areas are very specific, for instance, a particular crop plant. Collecting data in different sequences and directions may cause delay, error, or conflict with the system if not well designed and planned.
- **Collaborative communication**: A significant part of CCP-CPS is collaborative communication among the agents. In most cases, a mobile robot receives data

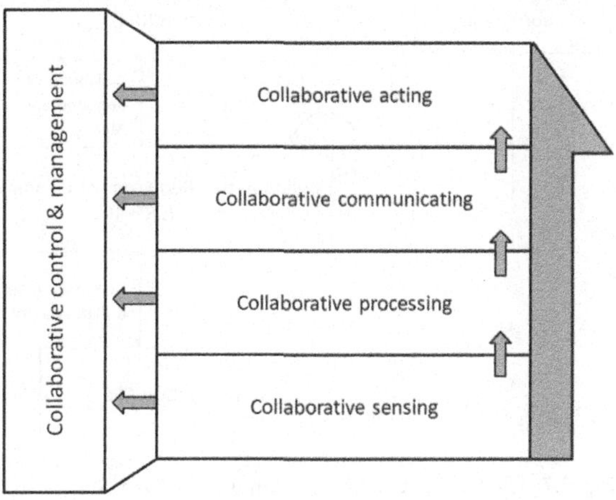

Fig. 3.6 Collaborative architecture for CCP-CPS

from sensors mounted on it. All data is combined and transmitted to a human with the expert system in order to diagnose crop plant disease in a timely manner.

- **Collaborative acting**: Acting effectively and optimally based on the timely outcomes from the previous three levels necessarily involves the coordination of humans, a robot, and sensors. In order to effectively monitor, detect, and respond promptly to the current status of crop plants, collaborative acting is crucial for allowing real-time control and precise response to the identified stressed crop plants.
- **Collaborative control and management**: Lastly, the collaborative control and management level is the integration of various control mechanisms and functions in harmony. It enables the system to perform specific, well-determined operations, efficiently utilize resources, provide secure operations, and control fault-tolerant mechanisms, as explained next.

3.1.3.2 Collaboration Requirement Plan (CRP)

For the workflow and referring to the CRP principle in CCT, two stages of CRP, CRP-I and CRP-II, are deployed in CCP-CPS. In CRP-I, a detailed plan is generated based on the work objectives and available resources. In CCP-CPS, CRP-I is utilized to assign and manage tasks and resources in CPS. The initial assignment of tasks to resources, which is the function of CRP-I, is presented in Fig. 3.7.

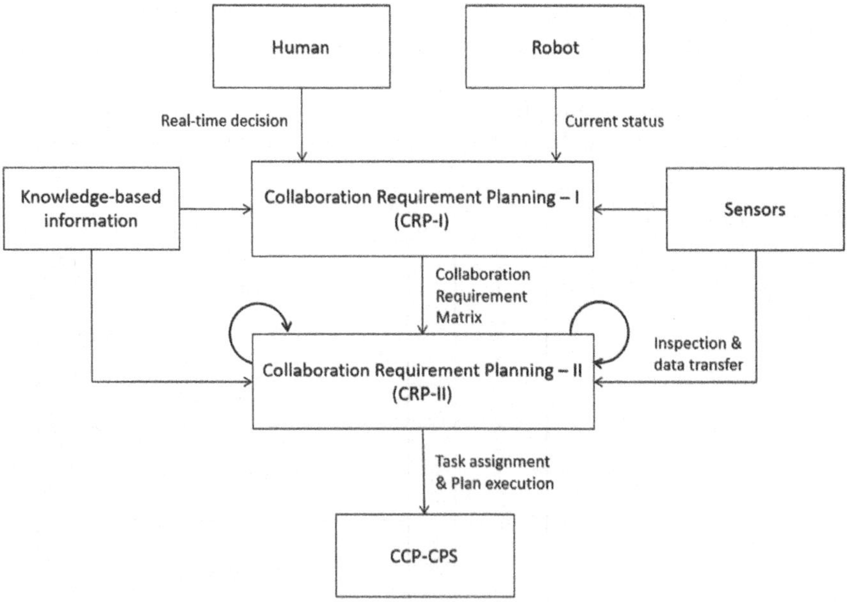

Fig. 3.7 CRP architecture for agricultural CCP-CPS

Moving to the second stage of CRP-II plan execution and revision, the objective is to adapt the plans determined from the CRP-I stage based on current conditions. Figure 3.7 shows an application of the CRP-II in the agricultural CPS. CRP-II executes the plan generated by CRP-I in real-time and revises the plan following and responding to current spatial and temporal challenges, changes, and constraints. The purpose of CRP-II is to assign and schedule (or re-assign and re-schedule) tasks to resources. CRP-II identifies all existing plans for each task, resolves the conflicts between the plans, and supports collaboration within a plan.

3.1.3.3 Error Prevention and Conflict Resolution (EPCR)

As its name implies, the EPCR principle deals with cyber-supported detection of Conflicts and Errors (C&E) among collaborating agents, as well as the cost associated with resolving the detected C&E. An ***Error*** is defined by CCT as a deviation from specifications' tolerance within or by a single agent or sub-system. A ***Conflict*** is defined by CCT as the inconsistencies between two or more agents or sub-systems. Naturally, any system that is unable to resolve its errors and disputes effectively will spin out of control and inevitably collapse. According to the definition and specification of C&E, they can occur at any agent in the scope of the CCP-CPS. For this case, C&E by humans, a robot, and sensors are primarily considered.

For illustration, C&E caused or influenced by a robot can be illustrated by the following example. Suppose the robot navigates at a greenhouse at 10 a.m. according to a scheduled routine. During the process, the robot approaches the assigned location, and sensors begin collecting data. The robot, however, is suddenly commanded to move to the next location before the sensors can complete the ongoing collection process. As a result, the data is found to be incomplete.

Continuing with this example, let us define a robot movement task as $Task_1$, and data collection task as $Task_2$, and two agents are matched to perform the data collection process. The conflict in this situation can be presented as follows.

$$\exists(Task_1(t) \cap Task_2(t)) \overset{agent\ performs>1\ task}{\rightarrow} C[ARS_{Group}(t)]$$

When a conflict occurred in the ARS at multiple agents at time t (represented by $ARS_{group}(t)$), first, there should be a mechanism to detect the conflict; for instance, the robot moves away and interrupts the ongoing data collection task by sensors. To resolve the conflict in this scenario, the following steps are specified, and the result is presented in Fig. 3.8.

Error Prevention and Conflict Resolution Algorithm	
1	**Detection**: A control module on the robot calls an interrupt to report the conflict to the farmer or control center
2	**Identification**: According to the definition of conflict, a control center identifies a conflict

(continued)

(continued)

3	**Diagnostics**: The control center determines the type, magnitude, time, and cause of the out-of-control status
4	**Control**: The control center analyzes, predicts, and prevents the propagation of the conflict
5	**Conflict resolution**: The control center resolves the conflict with a method that defines each and every event priority Note that an example of Step 5 is a mechanism to set the event's priorities. Define regulations of priorities as follows: – Regular events' priorities are lower than emergency events – Earlier emergency events' priorities are higher than later emergency events' priorities. (It can be changed with emergency priority code if needed.) – A robot can execute only one event task (order) at time *t*
6	**Exception handling**: Managing exceptions, i.e., constructive (necessary) deviations from the process In this scenario, the robot will first respond to the emergency event (new order), even if pre-scheduled navigation or operation is in progress. Figure 3.8 shows the process

Note that preempted tasks (interrupted before completion) will have to be rescheduled for completion after the emergency event has been resolved.

3.1.4 Monitoring Alternatives

One way to assess the performance of the designed protocol is to compare it against the performance based on current practices. Assume that two schemes for monitoring greenhouse crop plants are:

(1) CPS scheme for greenhouse crop plant monitoring (designed protocol);
(2) Non-CPS scheme for greenhouse crop plant monitoring (current practice).

Since CPS enables real-time control and communication among system agents, the first scheme, the CCP-CPS design, can include humans, robots, and sensors. In contrast, the current practice can only have a mobile robot installed with several sensors and no human in the design (no "Human-in-the-loop" design). This is due to the inability of the Non-CPS scheme in connectivity, communication, and collaboration between humans, a robot, and sensors. Next, each scheme is described in detail.

3.1.4.1 CPS Scheme

Three main agents (humans, robots, and sensors) collaborate as a system in the CPS scheme. A mobile robot with hyperspectral (or multispectral) sensors travels to specified locations to examine crop plants. The CPS scheme's workflow of operation procedure and operation time of each step (in parentheses) are presented (Fig. 3.9a).

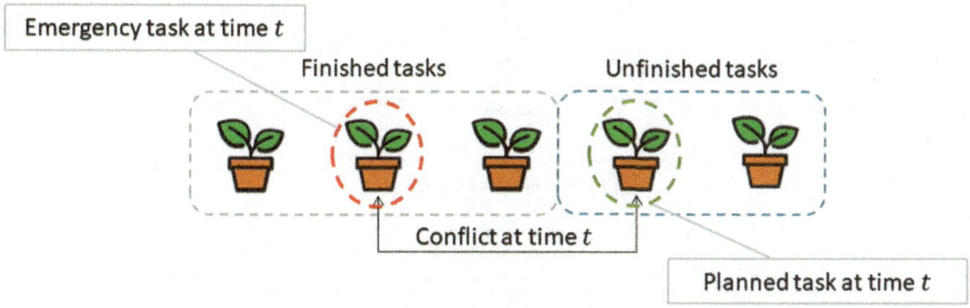

Fig. 3.8 Robot conflict propagation and the resolution process

As illustrated in Fig. 3.9b, the monitoring process by the CPS scheme begins at a base station where human operators work, and a mobile robot is based and begins its travels. The following event times are defined:

t_1—The robot travel takes time (t_1) to reach the assigned location.
At the location, the inspection process is performed along.
t_2, t_3, t_4, and t_5—Times used for localization, obtaining images, moving manipulator, and sensors operation, respectively.
t_6—Time required for calculating parameters and transmitting signals to human operators. t_6 also includes time for analysis by the human operators. In addition, this time may be further extended if the human operators decide to delegate the robot to re-measure specific parameters (in case C&E occurs during the operation). For instance, additional images or current temperatures at a given location are needed for a better decision or conclusion about the exact status of the given crop plants.

Otherwise, the robot goes on to the next location, and the process is repeated until the robot has visited all the assigned locations.

3.1.4.2 Non-CPS Scheme

In the Non-CPS scheme, as commonly applied, the system consists of a mobile robot and several sensors that are mounted on the robot. Figure 3.10a describes the Non-CPS scheme workflow and its operation time (in parentheses). Figure 3.10b depicts a pictorial representation of the scheme. The monitoring process begins at a base station, as it does with the CPS system.

Similarly, it takes t_1 for the robot to move to the assigned location under the same conditions. The robot and sensors require the same amount of time to perform the inspection process; t_2, t_3, t_4, and t_5.

Since there is no CPS facilitating in this system, it means that no real-time response about data quality or adequacy is possible. Hence, there is no direct online response from human operators. In this situation, the robot will not transmit data to the base station, and the robot will immediately travel to the next location. The process will

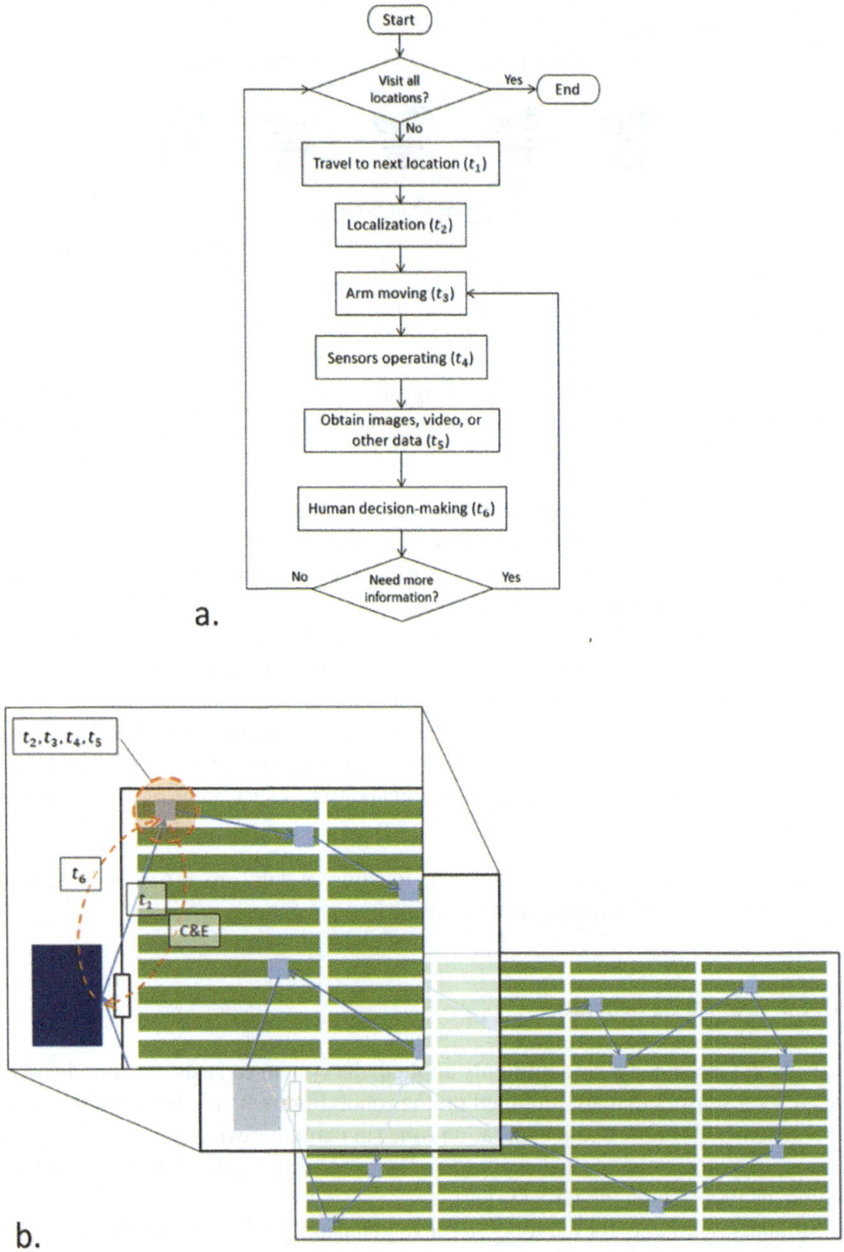

Fig. 3.9 CPS Scheme; **a** Workflow details; **b** Robot travel to examine and inspect crop plants

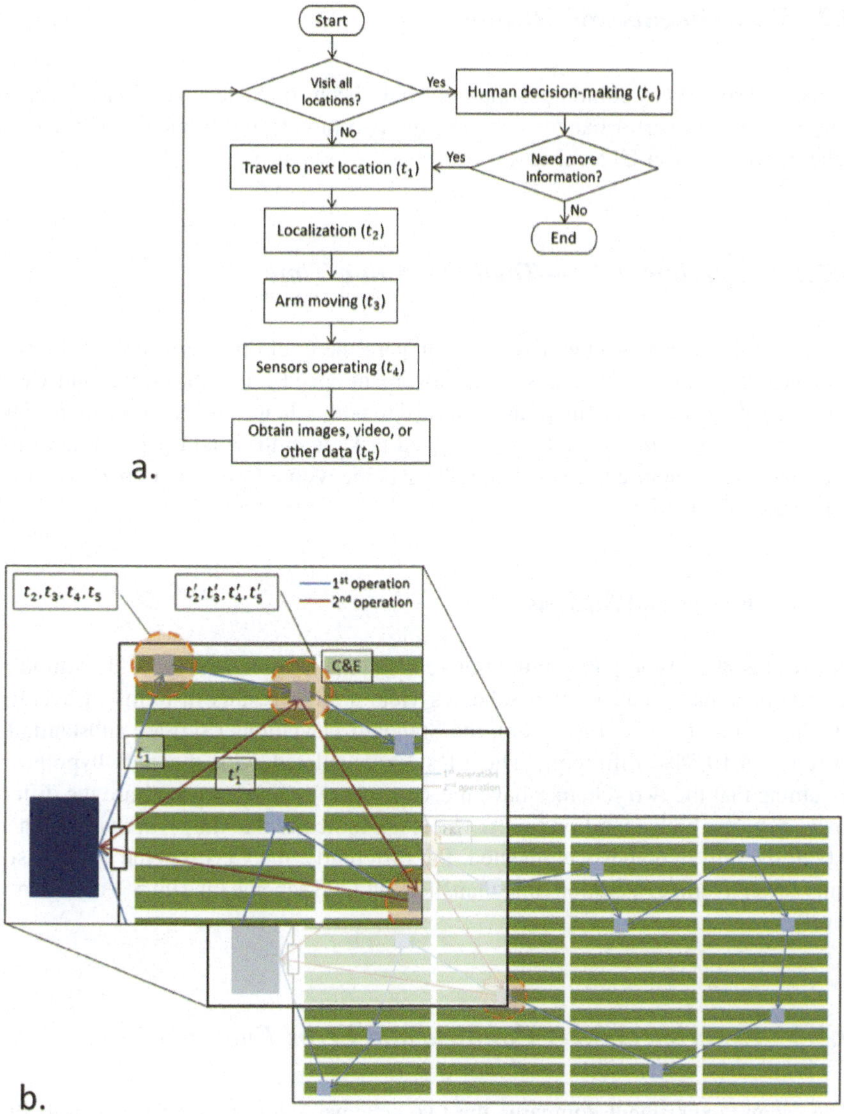

Fig. 3.10 Non-CPS Scheme. **a** Workflow details; **b** Robot travel to examine and inspect crop plants

be repeated until all locations have been visited. Then, the robot returns to the base station. Arriving at the base station, data will be transferred to a computer. If there is any additional information required, the robot will be dispatched again to obtain the missing data from the specific locations, which will take time to re-operate as t'_1, t'_2, t'_3, t'_4, and t'_5.

3.2 Experiments and Results

Three experiments by computer simulation have been conducted to evaluate different perspectives and performance criteria of the two monitoring frameworks, the CPS scheme and the Non-CPS scheme.

3.2.1 Experiment 3.1—Total Operation Time

For the first experiment, the aim is to compare the total operation time of the two schemes. The same parameters (i.e., movement speed, inspection rate, and C&E rate), target locations, and maps are assigned to both schemes in the experiment. The two schemes need to complete the assigned tasks, and the total operation time for each scheme is captured. To evaluate, the scheme with a lower total operation time is considered superior.

3.2.1.1 Results and Analysis

The results of 100 computer simulation replications are given in Fig. 3.11. Although the findings show that the two schemes yield similar results in terms of average operating time (0.16% difference), the Standard Deviations (SD) are substantially different (810.34% different). The t-test is calculated with the null hypothesis assuming that the two schemes have the same operation time to compare the differences between both operation times. The result indicates that the *p-value* is less than 0.005; the null hypothesis is rejected. Put differently, the CPS scheme has statistically lower average operation time than the Non-CPS scheme (at a 99.5% confidence level).

3.2.2 Experiment 3.2—Conflicts and Error Tolerance

The second experiment compares the two schemes while varying C&E probabilities. Since C&E can be anticipated in the agricultural context, the scheme with lower sensitivity, meaning higher tolerance to C&E, will be considered as a superior scheme. Put differently, the scheme with higher C&E tolerance is preferred over the other. Having a higher C&E tolerance, the total operation time will increase at a lower rate. In the experiment, the two schemes are assigned the same map, tasks, and parameters and are given a probability of C&E ranging from 0% (C&E free, ideal system) to 90%, with a 10% increment step. Finally, the total operation time for both schemes is estimated and analyzed for this experiment.

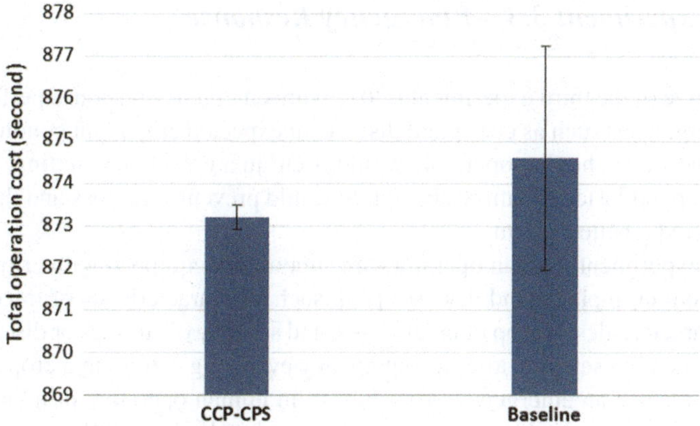

Fig. 3.11 Experiment 3.1 results

3.2.2.1 Results and Analysis

The results of 100 computer simulation experiments are shown in Fig. 3.12. As expected, the total operation time for both schemes increases when the system has a higher probability of C&E. The incremental rate of the operation time of the CPS scheme with increasing C&E probability, however, is lower than the operation time of the Non-CPS scheme.

To understand the statistical difference, the t-test is selected to compare the average operation time at each C&E level. As the *p-value* at each C&E level is less than 0.005, the CPS scheme has statistically lower operation time than the Non-CPS scheme at a 99.5% confidence level.

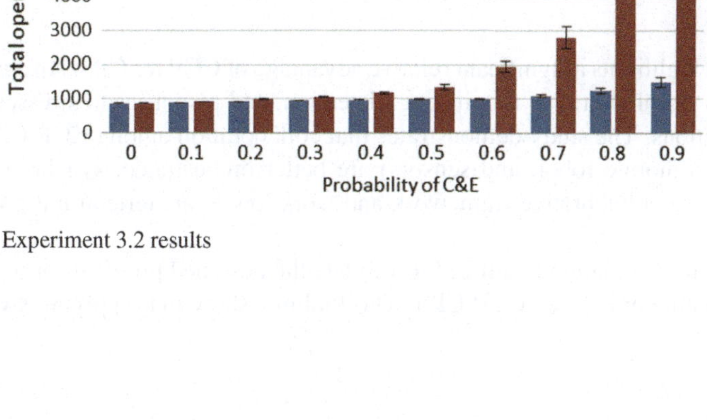

Fig. 3.12 Experiment 3.2 results

3.2.3 Experiment 3.3—Emergency Response

The objective of the third experiment is to capture emergency response performance. Since emergencies such as crop plant disease, unexpected crop plant conditions, and a new request from human operators would occur in a greenhouse setting, a scheme that can respond faster and more accurately would prevent yield loss and thus would be considered a better scheme.

In this experiment, human operators are supposedly equipped with critical information about crop plants and diseases [26], such as characteristics of infected crop plants, characteristics of crop plant diseases, and severity of stresses or diseases [14]. Hence, in the CPS scheme, after an emergency event, e.g., stress in a crop plant, the robot will receive an emergency procedure from human operators or a knowledge-based expert system to which the robot can immediately respond. In contrast, in the Non-CPS scheme, the robot has no specific procedure or guidance from human operators or an expert system. Therefore, the robot needs to randomly inspect crop plants surrounding the infected location to find the propagation direction before responding to the emergency.

3.2.3.1 Results and Analysis

The results from 100 computer simulation experiments are presented in Fig. 3.13. The average response time to an emergency and SD of the CPS scheme is significantly lower than the alternative. When CPS is applied, the robot focuses only on the direction of stress propagation provided by human operators or expert systems. On the other hand, without CPS, the robot needs to inspect crop plants one-by-one, making it less efficient. As a consequence, the process takes longer to complete. In addition to the advantage of human operators' knowledge assumption, it appears that the CPS scheme not only saves time and cost of operation but it also supports an environmentally friendly crop plant production system by reducing the use of pesticides [10] as the system applies chemical elements only at the needed locations.

3.2.4 Conclusion and Discussion

The result highlights a significant relative advantage of CCP for CPS. An agricultural CPS aims at collaborative monitoring, detection, and responses to stresses at identified locations. The study demonstrates that collaboration among CCP-CPS agents (humans, a mobile robot, and sensors) are better orchestrated, synchronized, and useful with a collaborative framework and workflow in an agricultural greenhouse setting.

Computer simulation is utilized to validate the designed protocol for the integration and collaboration of CCP-CPS. The findings show that applying CCP to the

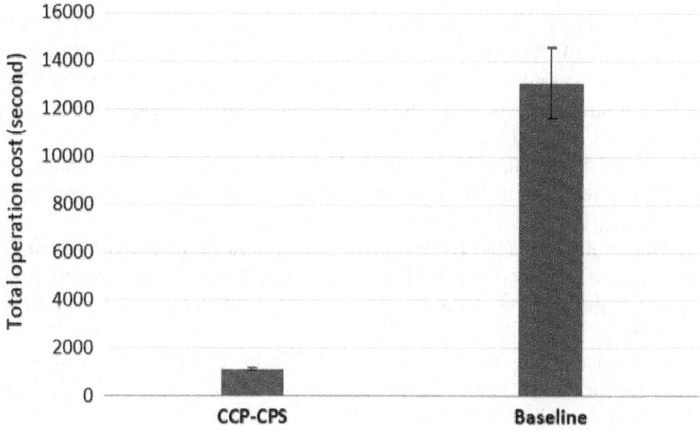

Fig. 3.13 Experiment 3.3 results

CPS environment can reduce the total operational time for monitoring and detection due to faster and more effective communication among participating agents. Furthermore, the system has a higher C&E tolerance. When C&E arise in a system, collaboration among system agents helps mitigate and stabilize the system's continuity, thus maintaining the system's operational time. Finally, when compared to an alternative system design, a CPS working under CCP-CPS will enable better emergency response. Therefore, we can conclude that system performance is significantly improved since collaboration among agents in the system is orchestrated, synchronized, and handled by CCP. While certain collaborative control features of the CCP and its relative advantages are clarified in this chapter, design details about agents, algorithms, and information flow applied by the CCP are explained in the next chapter.

References

1. Airlangga G, Liu A (2019) Initial machine learning framework development of agriculture cyber physical systems. In: Journal of physics: conference series, p 012065. https://doi.org/10.1088/1742-6596/1196/1/012065
2. Ajidarma P, Nof SY (2021) Collaborative detection and prevention of errors and conflicts in an agricultural robotic system. Stud Inform Control 30(1):19–28. https://doi.org/10.24846/v30i1y202102
3. Arampatzis T, Lygeros J, Manesis S (2005) A survey of applications of wireless sensors and wireless sensor networks. In: Proceedings of the 2005 IEEE international symposium on, mediterrean conference on control and automation intelligent control, pp 719–724. https://doi.org/10.1109/.2005.1467103
4. Behmann J, Mahlein A-K, Rumpf T, Römer C, Plümer L (2015) A review of advanced machine learning methods for the detection of biotic stress in precision crop protection. Precis Agric 16(3):239–260. https://doi.org/10.1007/s11119-014-9372-7

5. Chen S-L (2015) A power-efficient adaptive fuzzy resolution control system for wireless body sensor networks. IEEE Access 3:743–751. https://doi.org/10.1109/ACCESS.2015.2437897
6. Dimitriadis S, Goumopoulos C (2008) Applying machine learning to extract new knowledge in precision agriculture applications. In: 2008 panhellenic conference on informatics, pp 100–104. https://doi.org/10.1109/PCI.2008.30
7. Dusadeerungsikul PO, Liakos V, Morari F, Nof SY, Bechar A (2020) Smart action. In In: Agricultural internet of things and decision support for precision smart farming, pp 225–277. https://doi.org/10.1016/B978-0-12-818373-1.00005-6
8. Ehyaie A, Hashemi M, Khadivi P (2009) Using relay network to increase life time in wireless body area sensor networks. In: 2009 IEEE international symposium on a world of wireless, mobile and multimedia networks & workshops, pp 1–6. https://doi.org/10.1109/WOWMOM.2009.5282405
9. Finkelshtain R, Bechar A, Yovel Y, Kósa G (2017) Investigation and analysis of an ultrasonic sensor for specific yield assessment and greenhouse features identification. Precis Agric 18(6):916–931. https://doi.org/10.1007/s11119-016-9479-0
10. Gebbers R, Adamchuk VI (2010) Precision agriculture and food security. Science 327(5967):828–831. https://doi.org/10.1126/science.1183899
11. Goap A, Sharma D, Shukla AK, Rama Krishna C (2018) An IoT based smart irrigation management system using Machine learning and open source technologies. Comput Electron Agric 155:41–49. https://doi.org/10.1016/j.compag.2018.09.040
12. Guo P, Dusadeerungsikul PO, Nof SY (2018) Agricultural cyber physical system collaboration for greenhouse stress management. Comput Electron Agric 150:439–454. https://doi.org/10.1016/j.compag.2018.05.022
13. Huang C-Y, Yoon SW, Nof SY (2023) Brief history of the PRISM center and the PRISM global research network (PGRN). In: Huang C-Y, Yoon SW (eds) Systems collaboration and integration see past and future research through the PRISM center. Springer ACES, pp 3–50
14. Mahlein A-K (2016) Plant disease detection by imaging sensors—parallels and specific demands for precision agriculture and plant phenotyping. Plant Dis 100(2):241–251. https://doi.org/10.1094/PDIS-03-15-0340-FE
15. Moghadam P, Ward D, Goan E, Jayawardena S, Sikka P, Hernandez E (2017) Plant disease detection using hyperspectral imaging. In: 2017 international conference on digital image computing: techniques and applications (DICTA), pp 1–8. https://doi.org/10.1109/DICTA.2017.8227476
16. Nof SY, Ceroni J, Jeong W, Moghaddam M (2015) Revolutionizing collaboration through e-Work, e-Business, and e-Service, 1st ed
17. Pandey A, Tiwary P, Kumar S, Das SK (2019) A hybrid classifier approach to multivariate sensor data for climate smart agriculture cyber-physical systems. In: Proceedings of the 20th international conference on distributed computing and networking, pp 337–341. https://doi.org/10.1145/3288599.3288621
18. Samy I, Postlethwaite I, Gu D-W (2011) Survey and application of sensor fault detection and isolation schemes. Control Eng Pract 19(7):658–674. https://doi.org/10.1016/j.conengprac.2011.03.002
19. Schuster R, Schulter S, Poier G, Hirzer M, Birchbauer J, Roth PM, Bischof H, Winter M, Schallauer P (2011) Multi-cue learning and visualization of unusual events. In: 2011 IEEE international conference on computer vision workshops (ICCV workshops), pp 1933–1940. https://doi.org/10.1109/ICCVW.2011.6130485
20. Taki M, Abdanan Mehdizadeh S, Rohani A, Rahnama M, Rahmati-Joneidabad M (2018) Applied machine learning in greenhouse simulation; new application and analysis. Inf Process Agric 5(2):253–268. https://doi.org/10.1016/j.inpa.2018.01.003
21. Wang D, Vinson R, Holmes M, Seibel G, Bechar A, Nof S, Tao Y (2019) Early detection of tomato spotted wilt virus by hyperspectral imaging and outlier removal auxiliary classifier generative adversarial nets (OR-AC-GAN). Sci Rep 9(1):4377. https://doi.org/10.1038/s41598-019-40066-y

22. Wang L, Wang G (2016) Big data in cyber-physical systems, digital manufacturing and Industry 4.0. Int J Eng Manuf 6(4):1–8. https://doi.org/10.5815/ijem.2016.04.01
23. Wang Y, Liu H, Krogmeier J, Reibman A, Buckmaster D (2020) ISOBlue HD: an open-source platform for collecting context-rich agricultural machinery datasets. Sensors 20(20):5768. https://doi.org/10.3390/s20205768
24. Wani H, Ashtankar N (2017) An appropriate model predicting pest/diseases of crops using machine learning algorithms. In: 2017 4th international conference on advanced computing and communication systems (ICACCS), pp 1–4. https://doi.org/10.1109/ICACCS.2017.8014714
25. Welte JT, Ault AC, Bowman C, Ellis S, Buckmaster DR, Ess D, Krogmeier JV (2013) An approach to farm management information systems using task-specific, collaborative mobile apps and cloud storage services. In: American society of agricultural and biological engineers, Kansas City, Missouri, July 21–24
26. Wspanialy P, Moussa M (2020) A detection and severity estimation system for generic diseases of tomato greenhouse plants. Comput Electron Agric 178:105701. https://doi.org/10.1016/j.compag.2020.105701
27. Xu J, Gu B, Tian G (2022) Review of agricultural IoT technology. Artif Intell Agric 6:10–22. https://doi.org/10.1016/j.aiia.2022.01.001
28. Zervopoulos A, Tsipis A, Alvanou AG, Bezas K, Papamichail A, Vergis S, Stylidou A, Tsoumanis G, Komianos V, Koufoudakis G, Oikonomou K (2020) Wireless sensor network synchronization for precision agriculture applications. Agriculture 10(3):89. https://doi.org/10.3390/agriculture10030089

Chapter 4
Cyber-Collaborative Protocol for Early Detection of Stress in Crop Plants (CCP-ED)

Abstract To further explain the design of the Cyber-Collaborative Protocols (CCP), this chapter focuses on the design and application of CCP at the agent level, managing multiple agents, algorithms, and information flows. The objective of this chapter is to present the Cyber-Collaborative Protocol for Early Detection of Stress in Crop Plants (CCP-ED), which is derived from the CCP framework. The CCP-ED is developed and validated for coordinating a multi-agent system to work effectively, even optimally. The design consists of three main agents (humans, a robot, and sensors) and three main algorithms: Routing algorithm, Adaptive Search algorithm, and Stress Detection algorithm (Dusadeerungsikul et al. [9]; Dusadeerungsikul and Nof [8]). The Routing algorithm routes the robot to the sampled locations with the objective of minimizing travel cost/time, enabling the system to spend more time on higher value tasks, i.e., searching for stress and other signals in crop plants. In addition, the Adaptive Search algorithm, is activated after the Stress Detection algorithm indicates a stress in the crop plant, and provides a guideline for the system to respond to the crop plant status. The response is planned by utilizing knowledge-based information that incorporates data from crop plant pathologists, the stress characteristics of the particular crop plant, the current season, and the geographical location. The validation of the CCP-ED is conducted using numerical experiments, and the findings demonstrate that the CCP-ED shows superior results in comparison to the existing practice in two aspects: (1) the number of existing stressed crop plants found, and (2) the system's overall efficiency.

4.1 Task Description

The Agricultural Robotic System (ARS) is responsible for monitoring the status of crop plants in greenhouses, as discussed in the system scope assumptions (Chap. 2). With the objective to improve system performance, humans, a robot, and sensors collaboratively operate as a team [20], with decision-makers (humans), a facilitator (robot), and inspectors (sensors, with knowledge base information). In addition, with

P. O. Dusadeerungsikul and S. Y. Nof, *Cyber-Collaborative Algorithms and Protocols*, Automation, Collaboration, & E-Services 15,
https://doi.org/10.1007/978-3-031-56196-2_4

a human-in-the-loop design, the system is simplified [14], yet expected to deliver superior performance for the complex system [13].

The system begins with sampling the locations to inspect, based on the sampling policy, and a robot is directed from the Human-robot base station (see Fig. 4.1) to each assigned location. The robot is responsible for moving in a greenhouse and bringing sensors with detection agents to inspect the crop plants. The robot's manipulator will approach the crop plant, realign its attitude to enable sensors to collect data about stress/disease signals from various parts of the crop plant. The sensed data will be analyzed to indicate whether or not the crop plant is under control (i.e., not stressed). Moreover, given the evidence that when a crop plant shows unusual status, the nearby crop plants may also experience the same issue [11]. Crop plant stress and disease are also more likely to spread to adjacent areas in predictable directions [5] due to the influence of sunlight and other light sources, airflow direction, and other factors [3, 19, 23]. Hence, with such scientific knowledge, it is worthwhile inspecting the crop plants surrounding the first crop plant identified as stressed, in certain specific directions, to evaluate the severity of stress in the nearby area.

Let's assume that Northern and Western directions are more likely for stress to propagate. The ARS should emphasize these specific directions once the first crop plant with stress has been identified. As a result, the search algorithm should further examine the surrounding crop plants in the given directions. Plus, the algorithm must be adaptable in response to new information discovered during the operation time (possibly also updated online by remote experts). The situation mentioned above is depicted in Fig. 4.1.

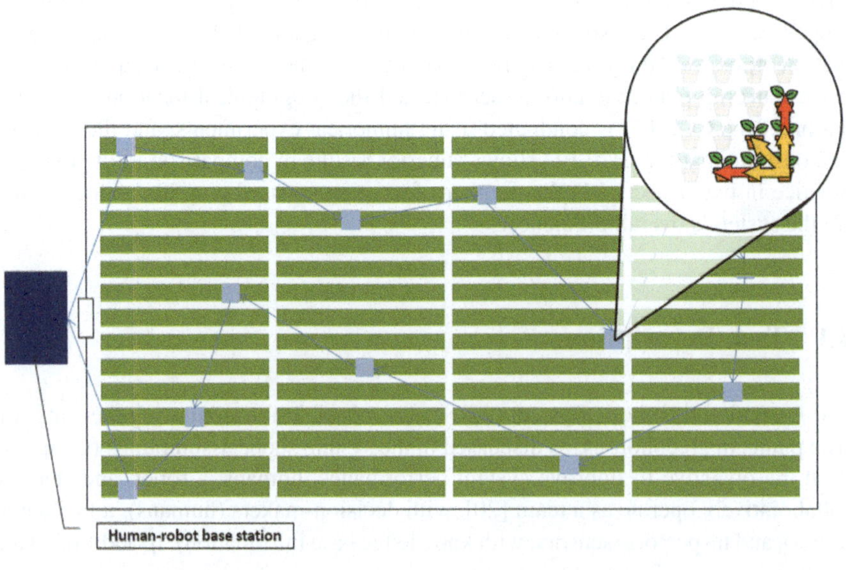

Fig. 4.1 ARS operation: North and West propagation directions assumed as high probability, as given and updated by experts for this crop plant season and geographical region

4.2 Cyber-Collaborative Protocol for Early Detection of Stress in Crop Plants (CCP-ED)

The Cyber-Collaborative Protocol for Early Detection of Stress in Crop Plants (CCP-ED) is based on the concepts of Cyber-Collaborative Protocol (CCP) and Collaborative Control Theory (CCT). CCT has been extensively utilized in various aspects to design multi-agent complex systems [15, 22, 29].

The system components are viewed as agents with the common goal of minimizing resources (time) used while identifying as many stressed crop plants as possible. To complete the task efficiently and effectively, the system agents must also cooperate intelligently [10, 12], utilizing CCP. Therefore, with CCP and CCT, CCP-ED is intended to incorporate agents in the ARS scheme to work collaboratively, harmoniously, and seamlessly.

The CCP-ED design process begins with the development of Collaboration Requirement Planning (CRP), which allows the designer to assign tasks to agents. Next, completing task preparation of possible Conflicts and Errors (C&E) in the system and protocol is addressed. Finally, the step-by-step procedure for harmoniously combining the system agents, algorithms, and information flow is explained.

4.2.1 Collaborative Control Theory (CCT) Principles in CCP-ED

4.2.1.1 Collaboration Requirement Planning (CRP)

The following section describes the CRP principle [28] used in CCP-ED. CRP consists of two parts, namely CRP-I and CRP-II. The first phase of the CRP process is referred to as CRP-I, which involves system planning. Conversely, CRP-II denotes the subsequent stage of execution and real-time modification within the CRP framework.

CRP-I: Plan Generation

The initial step of CRP is CRP-I, which is the planning step of the ARS. It is important to determine the requirements produced by the Collaboration Requirement Matrix (CRM) in order to develop the CRP-I. CRM can be created as follows.

$$Available(Resources_j) \times Task_i \rightarrow CRM \tag{4.1}$$

$Available(Resources_j)$ denotes the set of $Resources_j$ available for each $Task_i$. $Task_i$ denotes the tasks, and $Resources_j$ denotes the available agent or group of agents. The matrix will generate CRM with $CRM(Task_i, Resource_j) = 0$ when

Table 4.1a Task description

$Task_i$	
i	Task description
1	Compute the routing
2	Move to the location
3	Measure the parameter
4	Error and conflict checking
5	Stress status checking
6	Decision making

Table 4.1b Resource description

$Resource_j$	
j	Resource description
1	Robot
2	Sensors
3	Human
4	Team of robot and sensor
5	Team of robot and human
6	Team of sensors and human
7	Team of robot, sensors, and human

resource $Resource_j$ is not available for $Task_i$ and $CRM(Task_i, Resource_j) = 1$ when resource $Resource_j$ is available for $Task_i$.

For a given situation, the robot may inspect several parts of the crop plant at each location based on the information required. Moreover, in our designed ARS, only one robot is needed. (In other designs, a system could have more robots.) The system will have several types of $Task_i$ and $Resource_j$ under CCP-ED, as presented in Tables 4.1a, b. Also, note that the CRM can indicate alternative collaboration schemes under different conditions.

Therefore, the CRM, which has a group of agents in a row and a task in a column, is derived as follows to define what collaboration modes are feasible for a given system.

$$CRM = \begin{bmatrix} 0\,1\,0\,0\,0\,0 \\ 0\,0\,1\,0\,0\,0 \\ 1\,0\,0\,1\,1\,1 \\ 0\,1\,1\,0\,0\,0 \\ 1\,1\,0\,1\,1\,1 \\ 1\,0\,1\,1\,1\,1 \\ 1\,1\,1\,1\,1\,1 \end{bmatrix} \tag{4.2}$$

CRP-II: Plan Execution and Revision

CRP-II is the execution phase, which receives a generated plan from CRP-I. The CRP-II will update and revise the current plan when more real-time information is gathered during the operation. In this case, the initial plan is the route for the robot to inspect, together with the assigned agent to perform the task. When the ARS starts gathering information at each specific location, CRP-II will specify additional information (if needed) to be collected by sensors. Also, CRP-II will modify a mobile robot's route when unexpected events happen, modifying the initial plan from CRP-I. CRP-II will change the sequence of locations to be visited, the sequence of information to be collected at each location, and the assignment of sensor(s) to measure those parameters if necessary. Lastly, the plan will be revised over time based on new information discovered during the monitoring process, and the number of locations to visit might also be modified based on the remaining capacity of the robot.

4.2.1.2 Error Prevention and Conflict Resolution (EPCR)

Typically, there are possible C&E in the ARS because of the unstructured nature of the agricultural system, even though crop plants are grown in a greenhouse [4, 16]. By including C&E that may happen in the protocol and during operations, CCP-ED will account for real-life (not hypothetical) situations as well as potential challenges from applying CCP-ED. With C&E consideration, the protocol's true performance under realistic constraints during execution can be assessed [7].

EPCR principle of CCT will assist in resolving C&E as early as possible [17]. Errors occur when the input, output, or intermediate result of ARS does not meet specifications or standards. The term ***Error*** is described as follows.

$$\exists Error\left[ARS_{Single}(t)\right], if\, S_{ARS_{Single}}(t) \overset{Dissatisfy}{\rightarrow} N_r(t) \tag{4.3}$$

where $Error$ is an error in the system, $ARS_{Single}(t)$ is ARS single agent at time t, $S_{ARS_{Single}}(t)$ is the state of the ARS agent at time t, and $N_r(t)$ is the set of constraints, r, at time t.

Moreover, conflict refers to the difference between the information, goals, plans, tasks, operations, or activities of the collaborating agents. ***Conflict*** is defined as follows.

$$\exists Conflict\left[ARS_{Group}(t)\right], if\, S_{ARS_{Group}}(t) \overset{Dissatisfy}{\rightarrow} N_r(t) \tag{4.4}$$

where $Conflict$ is a conflict in system, $ARS_{Group}(t)$ is a group of ARS agents at time t, $S_{ARS_{Group}}(t)$ is the state of the group of ARS agents at time t, and $N_r(t)$ is the set of constraints, r, at time t.

Table 4.2 describes possible C&E in ARS. The C&E in a system should be resolved and/or prevented to maintain high system performance [1]. Therefore, if possible, each potential C&E in ARS must be addressed by specific prevention methods, including Machine Learning and Artificial Intelligence (ML&AI) techniques. In case C&E cannot be prevented, recovery procedures must be employed to mitigate their potentially resulting damage as soon as they occur. See further discussion of the EPCR of the CCT principle and its implementation in Chen and Nof [6], Reyes Levalle [18] and Zhong and Nof [27].

Table 4.2 Major potential error and conflict examples in ARS planning and control

Type	Potential example	Agent(s)	Situation	Consequence
Error	Input error	Human	Human inputs the wrong parameters into the system	Wrong decision and operation of robot and sensors
	Path error	Robot	Robot moves into the wrong path	Sensors and humans cannot receive the planned information
	Inspection error	Sensor	Wrong inspection or measuring the wrong parameters	Human makes the wrong decision, and robot might operate the wrong task
Conflict	Command conflict	Human, Robot	Human commands the robot to one location, but the robot moves to another location	Wrong location is inspected
	Information conflict	Human, Sensor	Information from sensors cannot be delivered to humans	Humans lack data to make precise decision
	Time conflict	Robot, Sensor	Robot starts moving to a new location, but the sensors do not complete the inspection	Data collected is incomplete
	Communication conflict	Human, Robot, Sensor	Communication among agents is interrupted	The synchronization among agents fails
	Sensor conflict	Two sensors	Two sensors deliver conflicting information	The interpretation of the information might not be accurate
	Decision conflict	Human and expert system	The decision from humans and the expert system conflicts	Humans may command the robot to collect more data

4.2.2 CCP-ED Design

This section, the CCP-ED step, is presented with pseudo-code. After specifying agent roles and responsibilities (by CRP) and C&E in the system, as well as how to prevent/resolve them (by EPCR), the CCP-ED is developed to synchronize agents, enabling the system to operate smoothly. The protocol includes three main algorithms: Routing algorithm, Adaptive Search algorithm, and Stress Detection algorithm. Figure 4.2 presents interconnection among algorithms under CCP-ED. Since the emphasis of this analysis is on the ARS collaborative protocol, the Stress Detection algorithm [24, 25] is used but not addressed here in detail.

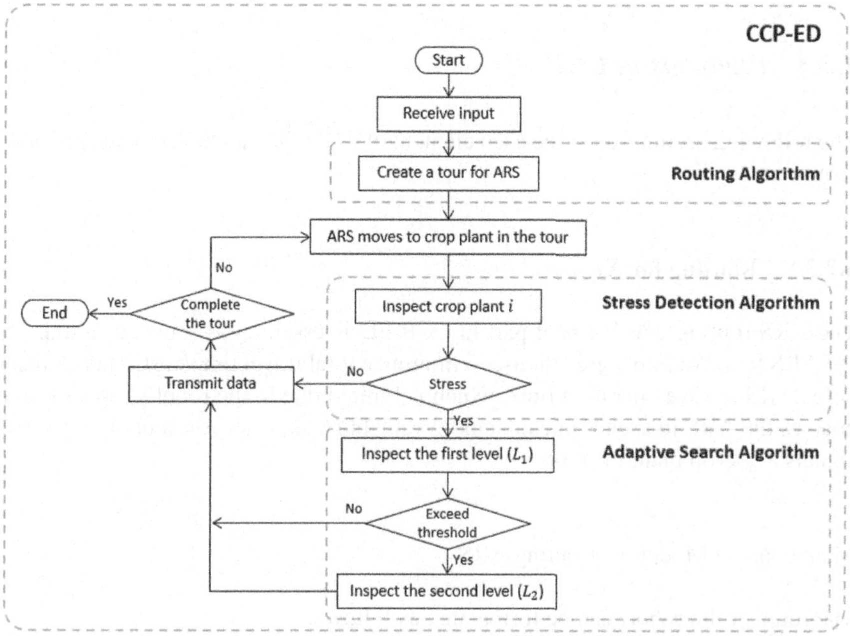

Fig. 4.2 Cyber-Collaborative Protocol for early detection of stress in crop plants (CCP-ED)

CCP-ED	
1	CCP-ED initialization
2	Routing algorithm creates a tour
3	**FOR** i from 1 to #location, **DO**
4	Stress Detection algorithm investigates location i
5	**IF** Stress Detection algorithm found stress, **DO**
6	**Activate** Adaptive Search algorithm

(continued)

(continued)

CCP-ED	
7	**ELSE**
8	**END IF**
9	**END FOR**
10	Protocol Terminated

Note that the role of the protocol is to activate (and terminate) the appropriate algorithm at the right time, when needed, and based on the result of the algorithm and other data inputs (from an intelligence agent such as the human or AI agent), decide how to proceed.

4.2.3 Elements in CCP-ED

The following sections describe each element of CCP-ED mentioned in the previous section in detail.

4.2.3.1 Routing for ARS

The ARS routing development part in CCP-ED is essential as it creates a tour for the ARS to move into a greenhouse to minimize total travel time/cost. This element helps ARS to save operation time, which is limited due to the robot's capacity, and enables the ARS to spend time on a more valuable task, such as searching for stress signals at a crop plant location.

Mathematical Model for Routing ARS

The route of the robots can be formulated as follows.

Decision variables:

$$x_{ij} = \begin{cases} 1\, if\ ARS\, traverse\ from\, i\, to\ j \\ 0\, otherwise \end{cases}$$

Parameters:

$$t_{ij} = Travel\, time\ from\, i\, to\ j$$

Objective function

$$\min z = \sum_{j=1}^{n} \sum_{i=1}^{n} t_{ij} x_{ij} \tag{4.5}$$

Subject to

$$\sum_{i} x_{ij} = 1 \; for \; all \; j \tag{4.6}$$

$$\sum_{j} x_{ij} = 1 \; for \; all \; i \tag{4.7}$$

$$\sum_{j \in G\prime} \sum_{i \in G} x_{ij} \geq 2 \; for \; all \; i \; and \; j \tag{4.8}$$

$$x_{ij} \in \{0, 1\} for \; all \; i \; and \; j \tag{4.9}$$

The objective function (4.5) is to minimize the total cost of traveling for ARS. If $x_{ij} = 1$, the ARS will traverse from i to j.

Constraint (4.6) forces each location to have one ARS incoming visit. With the same logic, constraint (4.7) forces all locations to have one ARS outgoing visit. Constraint (4.8) is the sub-tour elimination constraint by letting G and $G\prime$ be the partition of locations $i = \{1, 2, \ldots, n\}$. The constraint will ensure that ARS traverses from G to $G\prime$ and vice versa. Lastly, constraint (4.9) sets decision variables as binary.

Routing Algorithm for ARS

The above mathematical model is an NP-hard problem that cannot be solved in polynomial time [26]. An effective Routing algorithm that creates an optimal or near-optimal tour within time limits is required to direct a mobile robot to assigned locations. An efficient Routing algorithm will reduce travel time for mobile robots, allowing the ARS system to focus more on finding infected crop plants. A genetic algorithm is used in this work to construct a tour for a mobile robot, as shown in the following algorithm.

Routing Algorithm for ARS	
1	Algorithm initializes
2	$n \leftarrow \#location \times 10$
3	**FOR** i from 1 to n, **DO**
4	$Route_i \leftarrow Randomgeneration$
5	$Fitted_i \leftarrow Totalcostofroutei$
6	**END FOR**
7	**FOR** j from 1 to m, DO

(continued)

(continued)

Routing Algorithm for ARS
Select m routes based on the roulette wheel
Single crossover
Mutation
FOR k from 1 to $m/2$, **DO**
$Fitted_k \leftarrow Totalcostofroutek$
END FOR
END FOR
Terminate algorithm

4.2.3.2 Adaptive Search for ARS

The Adaptive Search algorithm is an essential part of the CCP-ED to indicate the severity of stress at crop plants in a greenhouse. Stress usually spreads in scientifically predictable directions based on the behavior and given conditions (i.e., type of stress, type of crop plant, environmental conditions, etc.) [21, 24]. Therefore, with this knowledge, we can develop an Adaptive Search algorithm that reflects the actual characteristics of crop plant stress propagation.

To illustrate the example associated with the use of the Adaptive Search algorithm, assume the example of the Northern and Western directions of stress to most likely spread. Hence, in such directions in the greenhouse, crop plants are more likely to have similar stress symptoms [2]. Equipped with this knowledge, once the first crop plant with stress is identified (see Fig. 4.3), the Adaptive Search algorithm should be activated to further inspect crop plants in the given most likely spread directions.

Based on the information discussed above, the Adaptive Search algorithm steps and pseudo code are as follows:

Adaptive Search Algorithm for ARS
Algorithm initializes
Parameter
$i \leftarrow$ crop plant status
IF $i =$ stress, **DO**
Adaptive Search algorithm is activated at the first level (L_1)
$l \leftarrow$#stressed crop plant locations
IF $l >$ Threshold, **DO**
Adaptive Search algorithm is activated at the second level (L_2)
END IF
END IF

(continued)

(continued)

Adaptive Search Algorithm for ARS	
11	Information transmitted
12	Terminate algorithm

By performing all steps of the AS, farmers can understand the magnitude and number of stressed or already infected crop plants. Hence, the farmers can plan a precise and localized mitigation procedure. This technique would save time searching for stressed crop plants in unlikely locations.

4.3 Experiments

4.3.1 Experiment Setting

The CCP-ED is used in computer simulation to verify the protocol's effectiveness. The protocol is evaluated against alternative protocols given the same operational resources (time, agent capacity, and number of agents). The protocols used in the experiment are shown in Table 4.3.

4.3.2 Performance Metrics

The metrics listed below are utilized to determine the protocol's efficiency in various aspects.

4.3.2.1 Number of Infected Crop Plants Found (NICF)

The first metric is the total number of stressed crop plants' locations discovered, as presented by Eq. (4.10).

$$NICF = \sum_{i=1}^{SP} d_i \tag{4.10}$$

$NICF$ = Total number of infected crop plants found
d_i = Number of infected crop plants found at location i
SP = Number of sampled locations in the greenhouse

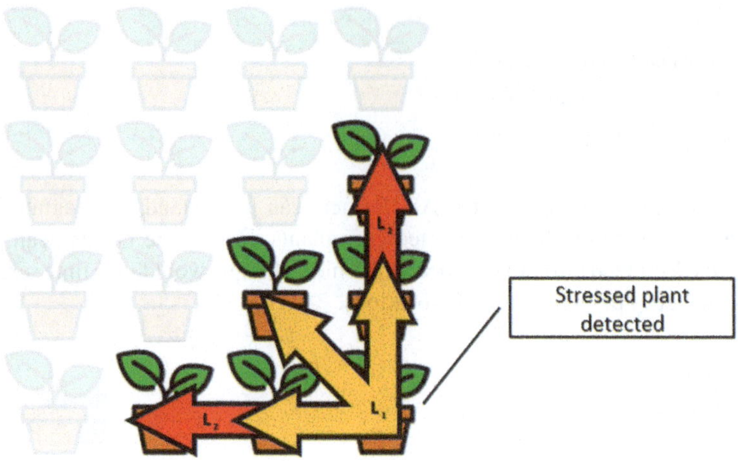

Fig. 4.3 Adaptive Search algorithm when propagation directions are known from previous pathological research

Table 4.3 Protocol used in the experiments

Protocol No	Routing algorithm	Search algorithm	Stress detection algorithm
1 (CCP-ED)	Routing algorithm for ARS	Adaptive Search	OR-AC-GAN
2	Routing algorithm for ARS	Always search	OR-AC-GAN
3	Routing algorithm for ARS	None search	OR-AC-GAN
4	Random routing	Adaptive Search	OR-AC-GAN
5	Random routing	Always search	OR-AC-GAN
6	Random routing	None search	OR-AC-GAN

4.3.2.2 Overall Robotic Effectiveness (ORE)

Overall Robotic Effectiveness (ORE) assesses the robotic system's overall detection ability. The measurement comprises three main components: Utilization (U), Performance (P), and Success (S). Each of the components measures a different aspect of the system.

- Utilization (U) measures the proportion of uptime in the total available time for a robot in the ARS system.
- Performance (P) measures the time that a robot performs work (searching for stressed crop plants) during its uptime.
- Success (S) measures the percentage of successful operations that have been completed by the robot, meaning the proportion of infected/stressed crop plants found out of the total number of crop plants inspected by this robot.

$$ORE = U \times P \times S \tag{4.11}$$

where

$$U = \frac{\sum_{j=1}^{SP\prime} \sum_{i=1}^{SP} (p_i + q_i + c_{ij})}{T} \tag{4.12}$$

$$P = \frac{\sum_{i=1}^{SP} (p_i + q_i)}{\sum_{j=1}^{SP\prime} \sum_{i=1}^{SP} (p_i + q_i + t_{ij})} \tag{4.13}$$

$$S = \frac{\sum_{i=1}^{SP} d_i}{\sum_{i=1}^{SP\prime} s_i} \tag{4.14}$$

U = Utilization
P = Performance
S = Success
T = Total available time
p_i = Inspection time at location i
q_i = Searching time at location i
t_{ij} = Travel time from i to j
d_i = Number of infected crop plants found during operation at location i
s_i = Number of inspected crop plants at location i
$SP, SP\prime$ = number of sampled locations in the greenhouse

The above metrics will be applied in the computer simulation experiments, which will be described in the following section.

4.3.3 Experiment Results

From 100 runs of computer simulation experiments, the results are presented in Figs. 4.4 and 4.5. The findings demonstrate that CCP-ED discovers a higher $NICF$ relative to alternatives. When compared to the baseline protocol (Protocol#6), the CCP-ED increases the $NICF$ by 73%.

To statistically test the difference in a mean of $NICF$, ANOVA tests are performed, and results are presented in Table 4.4. According to the findings, at least one protocol design produces a difference in mean of $NICF$ at the 0.005 significance level. Furthermore, the Fisher Pairwise comparison tests are used for a more in-depth study of protocol performance results. The results show that CCP-ED produces the highest $NICF$, which is one of the primary goals of ARS, by more effectively suggesting and identifying crop plants' stress locations in the greenhouse.

ORE is measured and evaluated to compare the efficiency of resource consumption, which is calculated from the system's U, P, and S [8] and presented in Fig. 4.5. ANOVA test indicates that, based on *p-value*, which less than 0.005, at least one protocol has a difference in ORE. Again, the Fisher Pairwise comparison tests are applied, and the results show that the CCP-ED has the highest ORE value. Because

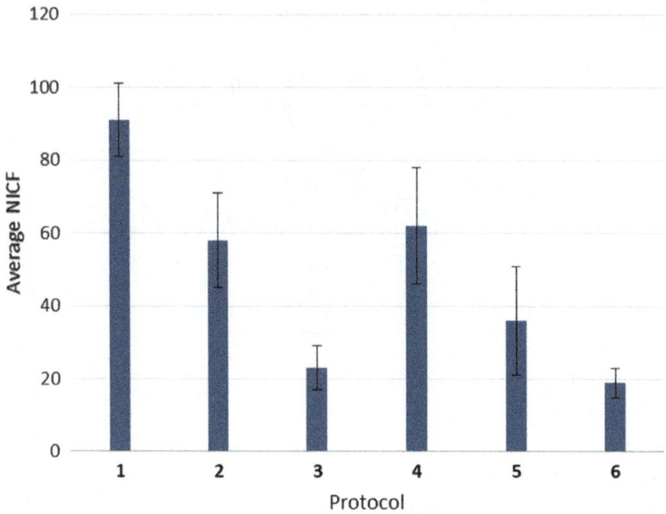

Fig. 4.4 $NICF$ under each protocol design

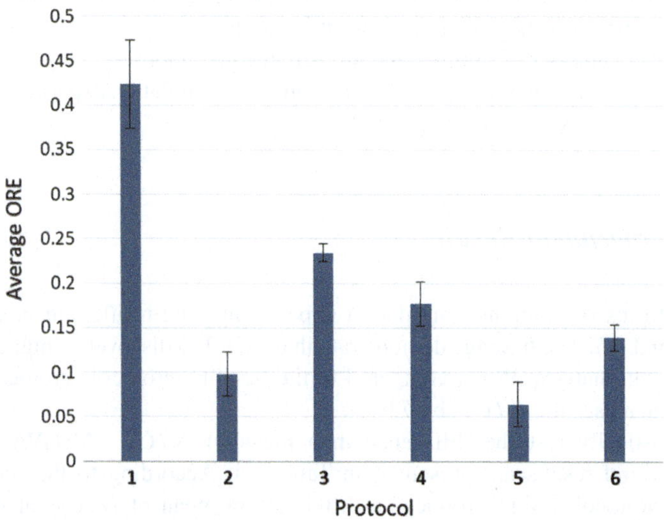

Fig. 4.5 ORE under each protocol design

of the highest ORE value ($ORE = 30.23\%$), the CCP-ED can be considered a protocol that can most effectively utilize a system's resource (time).

Table 4.4 ANOVA analysis

ANOVA analysis of the experiment results	*p-value*
H_0: The average of $NICF$ is the same among protocol designs H_a: The average of $NICF$ is not the same among protocol designs	<0.005
H_0: The average of ORE is the same among protocol designs H_a: The average of ORE is not the same among protocol designs	<0.005
Note: Because in both analyses *p-value* <0.005 it indicates that at 99.5% confidence level, at least one protocol has a difference in $NICF$ and ORE.	

4.3.3.1 Analysis of ORE

Figure 4.6 depicts ORE and its components, U, P, and S, averaging for each protocol after 100 simulation runs. The CCP-ED has the highest ORE among alternatives, suggesting that it has the highest efficiency of a monitoring system. The investigation of ORE shows that, despite the fact that protocols 3, 4, 5, and 6 have the highest utilization (U), the system stopped before visiting all nodes because it spent the majority of time on unnecessary tasks, such as traveling or searching the area that has no or minimal potential for stressed crop plants.

The CCP-ED outperforms others by detecting more existing crop plants with stress using the same resources. The protocol can direct the robot to visit all allocated locations before the system has been stopped because of depleted resources (time).

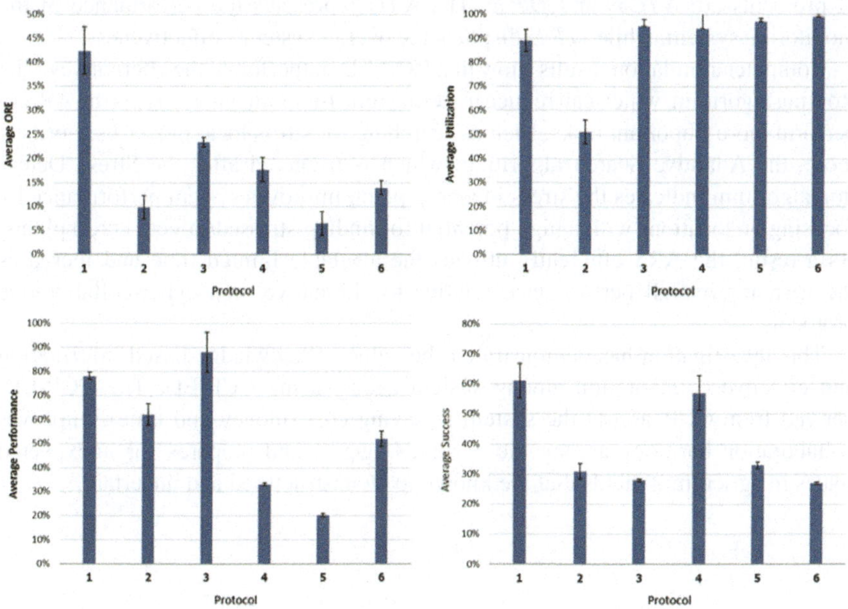

Fig. 4.6 ORE, utilization, performance, and success from each protocol

Moreover, the relatively better performance (P) implies that the CCP-ED uses the time to perform inspection tasks (rather than traveling). Finally, the high success rate (S) that CCP-ED achieved indicates that the Adaptive Search algorithm can successfully capture the characteristics of stress in crop plants.

4.4 Conclusion and Discussion

In this chapter, we have discussed the role of CCP at the agent level, using CCP-ED as an example. The CCP-ED has been established to improve the capability of the Collaborative Control Protocol for Cyber-Physical Systems (CCP-CPS) (which is described in the previous chapter). The CCP-ED aims to monitor the condition of greenhouse crop plants using a collaborative system-of-systems, combining humans, a mobile robot with manipulator, and sensors that function as a joint local agent and are not just individually involved in the CCP-CPS. The CCP-ED utilizes the CPS and Internet of Things/Internet of Services (IoT/IoS) as the connection and communication elements (see Chap. 3 for more details). The CCP-ED also relies on three algorithms, namely Routing algorithm, Adaptive Search algorithm, and Stress Detection algorithm, to perform crop plant monitoring tasks in the greenhouse.

To evaluate the performance of CCP-ED, computer simulation experiments have been conducted. The CCP-ED has been compared to other protocols to investigate its performance. Two metrics employed for capturing and assessing the performance of protocols are $NICF$ and ORE. The $NICF$ presents the performance of the monitoring system, while ORE displays the overall system's effectiveness.

Computer simulation results show that CCP-ED outperforms the alternatives. The Routing algorithm, which can reduce traveling time for the agent, instructs the ARS to perform more important tasks, such as searching for stress locations tasks. Furthermore, the Adaptive Search algorithm, which is triggered after the Stress Detection algorithm indicates the stress in a crop plant, improves system performance by focusing on locations with a high potential for finding stressed/infected crop plants. As a result, the ARS efficiently utilizes the available limited time and increases the system's overall performance relative to alternative, non cyber-collaborative protocols.

This investigation has demonstrated the value of knowledge-based information and effective collaboration among system agents using CCP-ED. The CCP-ED, derived from CCP, assists the system in saving cost (money and time), improves collaboration harmony among the system's agents, and prepares for unexpected issues in agricultural fields that are known to be unstructured and uncertain.

References

1. Ajidarma P, Nof SY (2021) Collaborative detection and prevention of errors and conflicts in an agricultural robotic system. Stud Inform Control 30(1):19–28
2. Asefpour Vakilian K, Massah J (2017) A farmer-assistant robot for nitrogen fertilizing management of greenhouse crops. Comput Electron Agric 139:153–163
3. Aylor DE (2003) Spread of plant disease on a continental scale: role of aerial dispersal of pathogens. Ecology 84(8):1989–1997
4. Bechar A, Vigneault C (2017) Agricultural robots for field operations. Part 2: operations and systems. Biosys Eng 153:110–128
5. Chanda M, Biswas M (2019) Plant disease identification and classification using back-propagation neural network with particle swarm optimization. In: 2019 3rd International conference on trends in electronics and informatics (ICOEI), pp 1029–1036
6. Chen XW, Nof SY (2012) Conflict and error prevention and detection in complex networks. Automatica 48(5):770–778
7. Duffuaa SO, Khan M (2005) Impact of inspection errors on the performance measures of a general repeat inspection plan. Int J Prod Res 43(23):4945–4967
8. Dusadeerungsikul PO, Nof SY (2019) A collaborative control protocol for agricultural robot routing with online adaptation. Comput Ind Eng 135
9. Dusadeerungsikul PO, Nof SY, Bechar A, Tao Y (2019) Collaborative control protocol for agricultural cyber-physical system. Procedia Manuf, ICPR-25 39:235–242
10. Feng F, Pang Y, Lodewijks G, Li W (2017) Collaborative framework of an intelligent agent system for efficient logistics transport planning. Comput Ind Eng 112:551–567
11. Golhani K., Balasundram SK, Vadamalai G, Pradhan, B (2018) A review of neural networks in plant disease detection using hyperspectral data. Inf Process Agric 5(3):354–371
12. Hoffman G, Breazeal C (2004). Collaboration in human-robot teams. In: AIAA 1st intelligent systems technical conference
13. John A, Mohan S, Vianny DMM (2021) Cognitive cyber-physical system applications. In: Cognitive engineering for next generation computing. Wiley, pp 167–187
14. Metcalfe JS, Perelman BS, Boothe DL, Mcdowell K (2021) Systemic oversimplification limits the potential for human-AI partnership. IEEE Access 9:70242–70260
15. Moghaddam M, Nof SY (2016) Real-time optimization and control mechanisms for collaborative demand and capacity sharing. Int J Prod Econ 171:495–506
16. Nair AS, Bechar A, Tao Y, Nof SY (2019) The HUB-CI model for telerobotics in greenhouse monitoring. Procedia Manuf, ICPR-25, 414–421
17. Nof SY, Morel G, Monostori L, Molina A, Filip F (2006) From plant and logistics control to multi-enterprise collaboration. Annu Rev Control 30(1):55–68
18. Reyes Levalle R (2018) Resilience by teaming in supply chains and networks, vol 5. Springer International Publishing
19. Roeber VM, Bajaj I, Rohde M, Schmülling T, Cortleven A (2021) Light acts as a stressor and influences abiotic and biotic stress responses in plants. Plant, Cell Environ 44(3):645–664
20. Rysz MW, Mehta SS (2021) A risk-averse optimization approach to human-robot collaboration in robotic fruit harvesting. Comput Electron Agric 182:106018
21. Sreeram M, Nof SY (2021) Human-in-the-loop: role in cyber physical agricultural systems. Int J Comput, Commun Control 16(2)
22. Velásquez JD, Nof SY (2009) Best-matching protocols for assembly in e-work networks. Int J Prod Econ 122(1):508–516
23. Velásquez AC, Castroverde CDM, He SY (2018) Plant-pathogen warfare under changing climate conditions. Curr Biol 28(10):R619–R634
24. Wang D, Vinson R, Holmes M, Seibel G, Bechar A, Nof SY, Luo Y, Tao Y (2018) Early tomato spotted wilt virus detection using hyperspectral imaging technique and outlier removal auxiliary classifier generative adversarial nets (OR-AC-GAN). 2018 Detroit, Michigan July 29–August 1

25. Wang D, Vinson R, Holmes M, Seibel G, Bechar A, Nof S, Tao Y (2019) Early Detection of tomato spotted wilt virus by hyperspectral imaging and outlier removal auxiliary classifier generative adversarial nets (OR-AC-GAN). Sci Rep 9:4377
26. Woeginger GJ (2003) Exact algorithms for NP-hard problems: a survey, pp 185–207
27. Zhong H, Nof SY (2020) In: Nof SY (ed) Dynamic lines of collaboration disruption handling & control, 1st edn
28. Zhong H, Nof SY, Berman S (2015) Asynchronous cooperation requirement planning with reconfigurable end-effectors. Robot Comput-Integr Manuf 34:95–104
29. Zhong H, Wachs JP, Nof SY (2013) HUB-CI model for collaborative telerobotics in manufacturing. Proc IFAC Intell Manuf Syst 46(7):63–68

Chapter 5
Cyber-Collaborative Optimal Search Protocol for Precision Agriculture (CCOSP)

Abstract In this chapter, we consider improvements in the Adaptive Search algorithm (AS) as examples of how Cyber-Collaborative Protocols (CCP) may increase algorithm performance, especially by minimizing the total cost of errors. The previous chapter has shown that AS has the potential to increase the monitoring system's efficiency. As a result, it is worthwhile to explore the algorithm further in order to determine the optimal AS procedure. The Cyber-Collaborative Optimal Search Protocol for Precision Agriculture (CCOSP), developed and analyzed in this chapter, considers the dynamics of a system, such as errors that may occur in the Agricultural Robotic System (ARS) and differences in crop plants' stress characteristics. CCOSP connects Dynamic Adaptive Search (D-AS) and the Stress Propagation Model (SPM), resulting in an optimal search procedure. CCOSP is developed, tested, validated, and integrated into the ARS to improve the system performance. Experiments are conducted to evaluate the CCOSP, and results show that CCOSP delivers the best performance compared to other search algorithms. Moreover, the sensitivity analyses show that CCOSP still yields high performance, even when the input parameter may deviate from the known true value.

5.1 Types of System Errors in a Monitoring System

System errors typically lead to suboptimal performance and waste system resources such as time and cost [1, 3, 11, 13].

In a crop plant monitoring system, there is the possibility of errors to happen [7]. In particular, the Stress Detection algorithm in the monitoring system may have two possible types of errors: Type 1 error (α) and Type 2 error (β). With an assumption of this study, the null hypothesis of the Stress Detection algorithm is that the crop plant does not have stress (H_0 : Crop plant is healthy). Hence, errors in the algorithm can be presented in Fig. 5.1, which depicts the relationships and circumstances of the system errors.

Type 1 error (α) or False Positive (FP) occurs when the Stress Detection algorithm, e.g., the algorithm developed by [21], indicates stress in the crop plant when, in fact,

P. O. Dusadeerungsikul and S. Y. Nof, *Cyber-Collaborative Algorithms and Protocols*, Automation, Collaboration, & E-Services 15,
https://doi.org/10.1007/978-3-031-56196-2_5

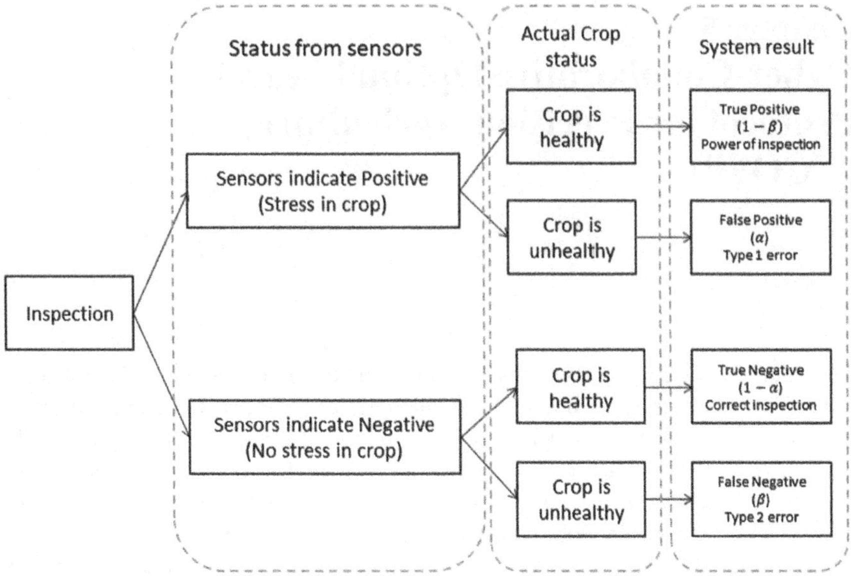

Fig. 5.1 Type of errors in a monitoring system

there is none [10]. When there is no error, the Stress Detection algorithm should show True Negative (TN). As a result, Eq. (5.1) presents the possibility of producing a type 1 error in the monitoring system.

$$\alpha = P(R|ST') \tag{5.1}$$

where

$R = Rejection\ region$

$ST' = Crop\ plant\ does\ not\ have\ stress$

Type 2 error (β) or False Negative (FN) is an error that occurs when the Stress Detection algorithm fails to detect stress in a crop plant when the crop plant actually has stress [10]. Without error, the Stress Detection algorithm must show a True Positive (TP). As a result, Eq. (5.2) shows the type 2 error in the monitoring system.

$$\beta = 1 - P(R|ST) \tag{5.2}$$

where

$R = Rejection\ region$

$ST = Crop\ plant\ has\ stress$

Furthermore, the power of the monitoring system, which is the ability to detect when there is stress in crop plants, is presented in Eq. (5.3)

$$Power\ of\ monitoring\ system = 1 - \beta = P(R|ST) \tag{5.3}$$

5.2 Impact of System Errors

To ensure crop plants' well-being and to prepare a treatment, if necessary, crop plants must be inspected [14]. Unfortunately, system errors in the monitoring process will result in information misinterpretation [23]. For example, a crop plant can be perceived as unhealthy and need further examination to determine the effect of stress or treatment when, in reality, the crop plant is healthy. In this scenario, time and resources are wasted, increasing the total operating cost [4, 18, 20].

To reduce operation costs and utilize resources efficiently, a system that can cope with errors by balancing the risk of error and the cost of an operation is needed [5, 17].

5.3 Monitoring Profit (MP)

During the monitoring process, the system will gain (or lose) system benefits [2]. Monitoring Profit (MP) is described as a benefit derived from the addition of new information to the system. The MP from discovering stress in a crop plant is denoted as MP_S, and the profit from discovering a healthy crop plant is denoted as MP_H.

Since crop plants with stresses need treatment to prevent a massive yield loss, the system will obtain more benefits when identifying crop plants with stresses than healthy crop plants [22],therefore, it is logical to conclude that $MP_S > MP_H$.

In addition, each step of the inspection will incur an inspection cost (C_I). That is, when the crop plant is healthy, the system should save operating costs (i.e., not operate), but if the crop plant contains stress, the system should inspect and indicate such location. As a result of this study, it is safe to say that $MP_S > C_I > MP_H$.

Furthermore, when errors (either type 1 or type 2) occur in the system, an additional cost may also incurred. To capture the effect of each error type, errors are quantified in terms of operating costs in the following section.

5.3.1 *Overage Cost (C_O)*

The overage cost (C_O) is an additional cost that occurs when the search algorithm spends time inspecting a location that is certain to have stress.

Appendix 5.1 Shows that type 1 error incurs the overage cost (C_O) and the amount of the overage cost is equal to $\alpha \times C_I$.

5.3.2 Underage Cost (C_U)

The underage cost (C_U) is an additional cost that arises when the search algorithm does not spend enough time inspecting the location where the stress information is not sufficiently clear.

In addition, Appendix 5.2 shows that the type 2 error incurs the underage cost (C_U) and the C_U is equal to $\beta \times C_I$.

5.4 Cyber-Collaborative Optimal Search Protocol for Precision Agriculture (CCOSP) Development

With the analysis of errors in the system in the previous section, the CCOSP can be developed to balance cost and benefit. In this section, Dynamic Adaptive Search (D-AS) is first developed by computing optimal search progression to maximize MP. Then, to support the D-AS, the Stress Propagation Model (SPM) is created.

5.4.1 Dynamic Adaptive Search Algorithm

In order to maximize MP, Dynamic Adaptive Search (D-AS) algorithm is developed. The critical ratio (R_0), which is required in D-AS can be calculated by Eq. (5.4). (Appendix 5.3 shows the derivation of R_0 by balancing C_O and C_U to reach the optimal MP.)

$$R_0 = \frac{\beta}{\beta + \alpha} \tag{5.4}$$

D-AS steps are presented below. The optimal expansion to maximize MP is when the search algorithm extends the search to the first m crop plants so that the cumulative density function (CDF), R, is less than or equal to the critical ratio, R_0. Note that the algorithm to calculate CDF of the stress propagation will be presented in the next section.

Dynamic Adaptive Search (D-AS) Algorithm	
1	Initialize
2	Parameter
3	Calculate Critical ratio (R_0)
4	**FOR** all locations m in monitoring plan, **DO**
5	Inspect m
6	**IF** m is stressed **DO**

(continued)

(continued)

Dynamic Adaptive Search (D-AS) Algorithm	
7	$R = R + P$ (the first crop plant has stress)
8	**WHILE** $R \leq R_0$ **DO**
9	$m = m + 1$
10	Inspect m
11	$R = R + P$ (m crop plant has stress)
12	**END WHILE**
13	**END IF**
14	**END FOR**
15	Terminate Algorithm

5.4.2 Cumulative Density Function for Stress Propagation

Since the D-AS includes the Cumulative Density Function (CDF) of the stressed crop plants, the algorithm for modeling and determining the probability of each location is required and presented in this section.

5.4.2.1 Stress Propagation Model (SPM)

Because of environmental conditions, the crop plant may generate stress on its own [12, 15]. Once the crop plant gets stress, the stress can propagate to nearby locations in some specific direction based on the type of stress, sunlight, seasons of the year, and airflow [9, 16].

Let p denote the probability that the crop plant develops stress, and q is the probability that stress propagates in a specific direction. The probability of stress propagation in a specific direction can be expressed by Eq. (5.5). (See Appendix 5.4 for the derivation of Eq. (5.5)).

$$P(\text{At most } m \text{ crop plants are stressed}) = \begin{cases} 1 - pq^m; \; m \geq 0 \\ 0; \; otherwise \end{cases} \tag{5.5}$$

Based on Equation (5.5), the Stress Propagation Model (SPM) can be utilized for generating CDF supporting D-AS.

Stress Propagation Model (SPM)	
1	Initialize
2	Parameter
3	$m \leftarrow$ Current location
4	Calculate P (Atmost m crop plants are stressed) $= P(M \leq m)$from Eq. (5.5)
5	Terminate Algorithm

5.4.3 *Cyber-Collaborative Optimal Search Protocol for Precision Agriculture (CCOSP)*

The CCOSP is presented in Fig. 5.2, which shows that the two algorithms (D-AS and SPM) operate in parallel, leveraging CCP connections as a facilitator for exchanging information. The CCOSP, which considers conflicts and errors within the system, employs adaptive search strategies that depend on the acquisition of new information relating to the status of each crop plant.

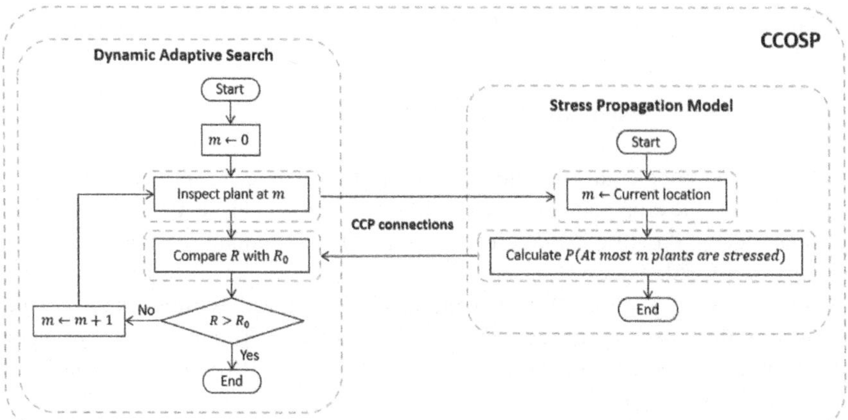

Fig. 5.2 Cyber-Collaborative Optimal Search Protocol for Precision Agriculture (CCOSP)

5.4.4 Alternative Search Algorithm

Apart from D-AS, three additional search algorithms can be utilized and are presented in this section. Therefore, a total of four search algorithms, including Dynamic-Adaptive Search, Static-Adaptive Search, Always Search, and No Search, are developed and applied in a greenhouse. Note that the first two search algorithms are classified as adaptive, while the last two are classified as non-adaptive. Three additional search algorithms (Static-Adaptive Search, Always Search, and No Search) are outlined in detail in the following sections.

5.4.4.1 Static Adaptive Search Algorithm

Static Adaptive Search (S-AS) is a variant of AS developed by [9]. The S-AS utilizes knowledge-based information to assist the algorithm in emphasizing the specific direction. Specifically, the knowledge-based information includes the stress propagation characteristics and directions to guide the search algorithm. Despite the fact that the algorithm will adjust based on new information discovered, the search progression is predetermined. The algorithm is presented as follows.

Static Adaptive Search (S-AS) Algorithm
Initialize
Parameter
FOR all location m in monitoring plan, **DO**
Inspect m
IF m is infected, **DO**
Inspect $m+1$ and $m+2$
Calculate Percentage of stress locations
IF Percentage of stress locations > predetermined threshold, **DO**
Inspect $m+3$ and $m+4$
END IF
END IF
END FOR

5.4.4.2 Always Search Algorithm

The Always Search algorithm represents the system in the absence of real-time information. The system agent is unable to determine whether or not the inspected location is stressed. The system, on the other hand, has scientific information about the stress propagation direction. As a result, in order to obtain the most information, the system will always continue the search in a specific direction. The Always Search algorithm is presented as follows.

Always Search Algorithm
Initialize
Parameter
FOR all location m in monitoring plan, DO
Inspect m, $m+1$, $m+2$, $m+3$, $m+4$
END FOR

5.4.4.3 No Search Algorithm

No search represents the current practice, which does not utilize knowledge-based information. Therefore, the worker will inspect only the assigned locations according to the given monitoring plan [8]. This logic is presented by the No Search algorithm, meaning there is no additional search whether or not a stress in the crop plant is detected. The No Search algorithm is presented as follows.

No Search Algorithm
Initialize
Parameter
FOR all location m in monitoring plan, **DO**
Inspect m
END FOR

Note that all algorithms presented in Sect. 5.4.4 (S-AS, Always Seach, and No Search) have no connections to other algorithms, so they are not considered cyber-collaborative, as CCOSP or CCP.

5.5 Experiments, Results, and Analysis

In this section, numerical experiments are conducted using computer simulation. The CCOSP is compared to the alternatives in terms of cost and system efficiency. Finally, a sensitivity analysis is performed to assess the robustness.

Table 5.1 Summary of parameter settings

Parameters	Value
Number of simulation runs	100
Type 1 error	α=10%
Type 2 error	β=30%
Monitoring Profit (when stress is found)	MP_S= 5 cost unit/location
Monitoring Profit (when stress is not found)	MP_H= 1 cost unit/location
Inspection cost	C_I= 3 cost unit/location
Search procedures	Four search procedures are applied in the experiments (CCOSP, S-AS, Always Search, and No Search)

5.5.1 *Experiment Setting*

The experiment begins with the inspection of 100 crop plants in the greenhouse, which are sampled randomly according to the environmental setting described in Chap. 2. Moreover, Table 5.1 summarizes the simulation parameters used in the experiments.

5.5.2 *Computational Experiments and Results*

The experiments have been conducted in this section to validate the developed protocol and algorithms.

5.5.2.1 System Performance Analysis

The system performance will be analyzed by using computer simulation experiments. The matrices utilized to capture system performance are True Positive Ratio (*TPR*) and True Negative Ratio (*TNR*).

- **True Positive Ratio**

The True Positive Ratio (*TPR*) indicates how many TP were discovered from the total positive (both TP and FP) location specified. The objective of the system is to indicate TP locations; therefore, the procedure (protocol or algorithm) that can deliver a higher *TPR* is considered a better option. *TPR* can be calculated as present in Eq. (5.6).

$$TPR = \frac{\sum_{i=1}^{M} TP_i}{\sum_{i=1}^{M} (TP_i + FP_i)} \quad (5.6)$$

where

M = Number of inspection locations.

TP_i= True Positive found at location i

FP_i = False Positive found at location i

- **True Negative Ratio**

True Negative Ratio (*TNR*) represents the ratio between TN and total negative location (both TN and FN). The procedure that delivers a higher *TNR* is the preferred option because it means that the procedure can indicate the true healthy crop plants. *TNR* can be calculated as presented in Eq. (5.7)

$$TPR = \frac{\sum_{i=1}^{M} TN_i}{\sum_{i=1}^{M} (TN_i + FN_i)} \tag{5.7}$$

where

M = Number of inspection locations.

TN_i= True Negative found at location i

FN_i = False Negative found at location i

Experiment Results

The results of the experiments are shown in Figs. 5.3 and 5.4. In general, the CCOSP outperforms (or performs as well as) others in terms of *TPR*. On the other hand, *TNR* results are very similar among different alternatives.

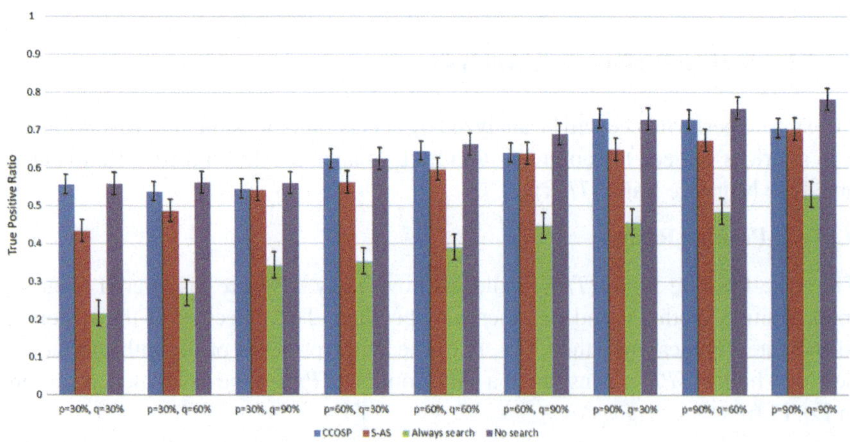

Fig. 5.3 Performance analysis (True Positive Ratio)

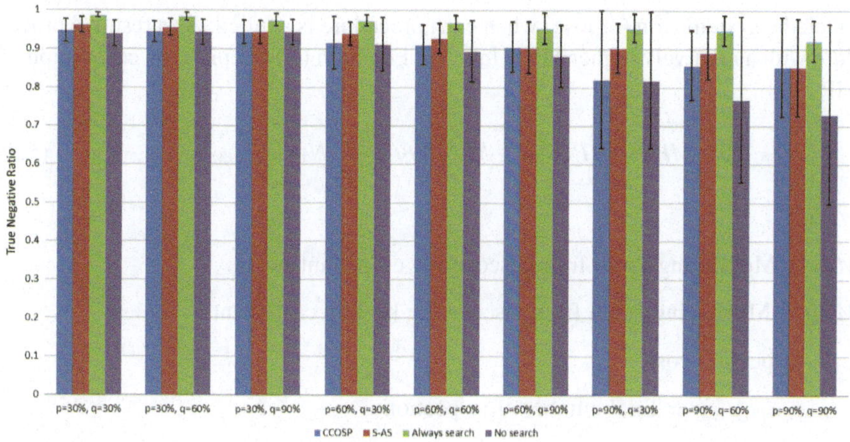

Fig. 5.4 Performance analysis (True Negative Ratio)

The ANOVA and post-hoc Bonferroni tests are used for further analysis. Results show that in the stresses with a low probability of developing or propagating, such as p or q equal to 30%, the D-AS in CCOSP performs as the No Search, and both outperform the other two algorithms with 95% confidence level. In contrast, the D-AS in CCOSP performs as well as an S-AS when stresses have a higher probability of generating or propagating (higher p and/or q).

With the Always Search, which is the most conservative approach, the *TPR* is lower than other algorithms. On the other hand, in terms of *TNR*, the Always Search outperforms other algorithms with a 95% confidence level.

To summarize, since the CCOSP has the ability to adjust itself based on new information, it modifies the search algorithm dynamically according to the ongoing inputs. For instance, if the inputs show a low propagation rate (q equal to 30%), the D-AS in CCOSP switches to the No Search. The D-AS, on the other hand, extends the search further if the propagation rate is high.

5.5.2.2 Cost Analysis

In this section, we analyze the CCOSP from cost perspective and compare it with the alternative procedures. The computer simulation experiment is used for analyzing each procedure, using the Monitoring Profit (MP).

- **Monitoring Profit (MP)**

Monitoring Profit (MP) represents the system's gain as a result of the new information from extending search in relation to the operation cost (as discussed in Sect. 5.3). The new information about the crop plant's status would lead to system benefits. On the other hand, it has an additional cost due to more inspections. The MP represents how much the system gain (or lose) when each procedure is applied. As a result, a procedure with a relatively higher MP is favored. Equation (5.8) depicts the calculation of MP.

$$MP = [MP_S N_{TP} + MP_H N_{TN}] - C_I N_{TP+FP+TM+FM} \tag{5.8}$$

where

MP_S = Monitoring Profit from discovering crop plant stress

MP_H = Monitoring Profit from discovering healthy crop plant

C_I= Inspection cost

$N_{TP+FP+TM+FM}$ = Total number of inspections.

Experiment Results

The results of the experiments are shown in Fig. 5.5. Comparing among the procedures, the CCOSP indicates the best performance. The ANOVA and Post-hoc Bonferroni tests are used to analyse the statistical differences. The results show, with a 95% confidence level, that the CCOSP is statistically significantly higher than other options in the majority of cases.

Only in the case where p and q are low (30%), the CCOSP and No Search result no statistical differences in terms of MP. The reason is that, in the low propagation rate case, the D-AS in CCOSP adapts itself to perform as No Search. It is reasonable because, as mentioned earlier, in the case that stress rarely spreads, extending search may not worth, even if the first location is found to be infected. The protocol should not expand the search to the near locations, because the cost of expanding will surpass the expected information gain.

The Always Search, on the other hand, has the lowest MP. The explanation for this result is that the algorithm examines additional locations regardless of the information received. Therefore, the algorithm costs the system heavily and yields the lowest MP.

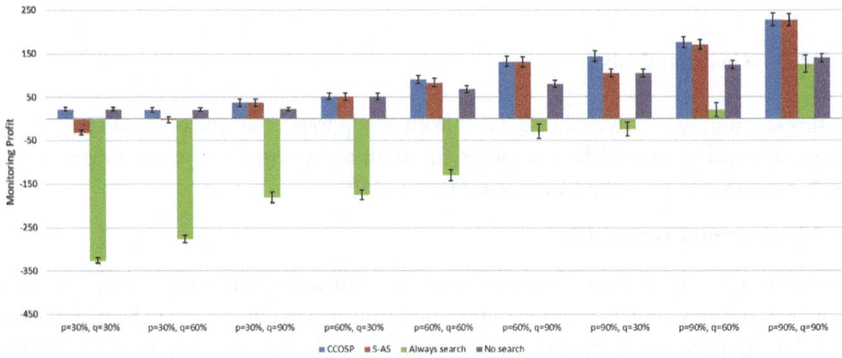

Fig. 5.5 Cost analysis (Monitoring Profit)

5.5.2.3 Sensitivity Analysis

When the conditions and parameters deviate from the assumptions in Table 5.1, how good can the algorithm be? To answer this question, the sensitivity of CCOSP is analyzed and compared to the alternatives. This section aims to examine the sensitivity and robustness of the proposed protocol and algorithms. Since the agricultural environment is less structured and has high variability and uncertainty, robustness is an important characteristic of a procedure that operates in it [19]. The procedure that is more robust (or less sensitive) to changes in the parameters is perceived to be a better option [6]. In this analysis, four parameters (namely, p, q, α and β) have been varied to observe the sensitivity of the results. The sensitivity can be defined as in Eq. (5.9).

$$Sensitivity = \frac{Y(x)}{Y(x^{*})} \tag{5.9}$$

where

$Y(x)$ = Result from assumed parameters.

$Y(x^{*})$ = Result from actual parameters (Optimal result).

Experiment Results

Sensitivity results from the simulation experiment are shown in Figs. 5.6, 5.7, 5.8 and 5.9. While Fig. 5.6 shows the sensitivities of search algorithms when p deviates from the known parameters, Fig. 5.7 shows the sensitivities when q has deviations. Figures 5.8 and 5.9 demonstrate the sensitivities when α and β have deviated from the settings. When the parameter deviations range from 10 to 25%, CCOSP yields results that are no more than 10% different from the optimal settings. In other words, the CCOSP is relatively robust and yields near-optimal results when the input is uncertain. Other procedures are less robust when compared to the CCOSP. The Always Search is the most sensitive, particularly when p and q are changed from the known value.

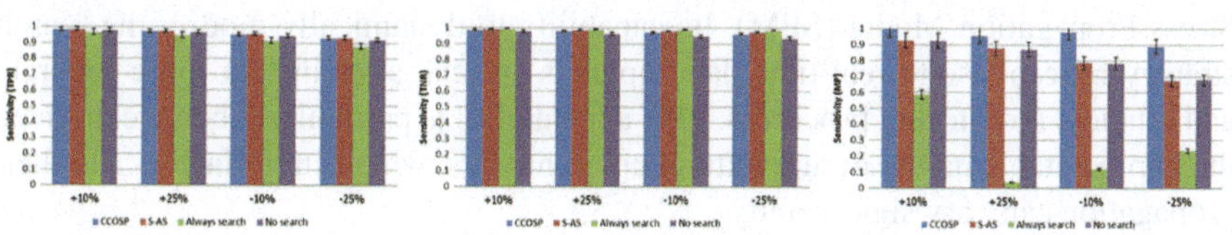

Fig. 5.6 Sensitivity analysis (when p value deviates from the known value)

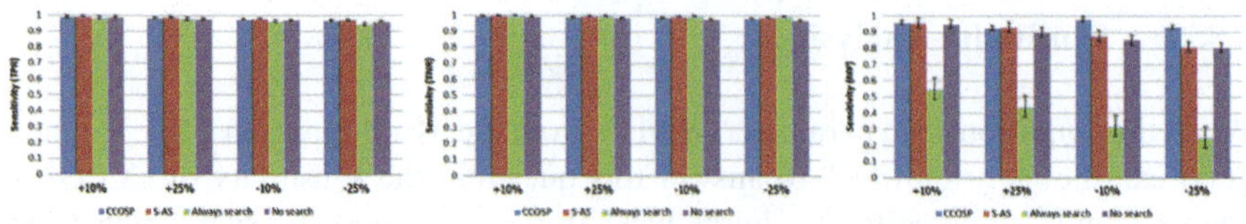

Fig. 5.7 Sensitivity analysis (when q deviates from the known value)

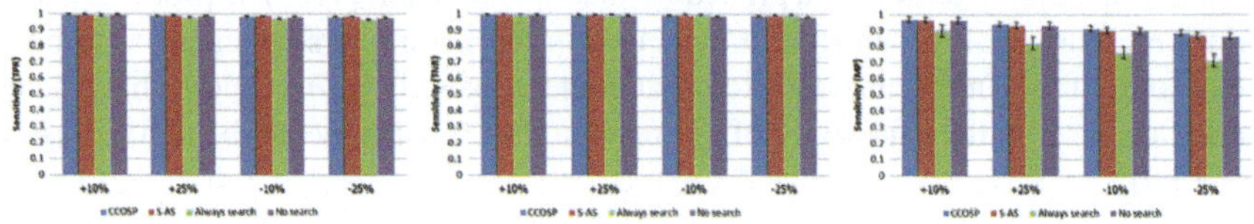

Fig. 5.8 Sensitivity analysis (when α deviates from the known value)

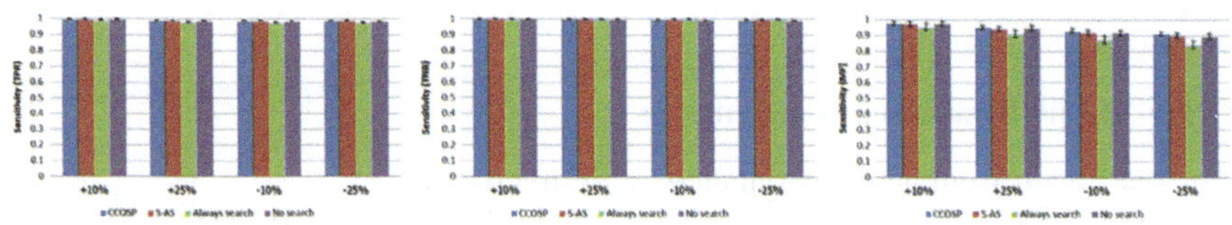

Fig. 5.9 Sensitivity analysis (when β deviates from the known value)

5.6 Conclusion and Discussion

This chapter presents the applications of CCP at the algorithm level. First, the study investigates system errors within crop plant monitoring systems. Type 1 and Type 2 errors, which can impact system performance and resource utilization, are defined and explained in this chapter. The Cyber-Collaborative Optimal Search Protocol for Precision Agriculture (CCOSP) is developed and evaluated to manage and provide optimal procedures in ARS to deal with system errors.

The CCOSP connects two algorithms: Dynamic Adaptive Search (D-AS) and Stress Propagation Model (SPM). Its capability to dynamically modify the search strategy in response to real-time data inputs is critical, allowing for more accurate and efficient monitoring processes. This adaptability is particularly beneficial given the unpredictable nature of agricultural environments, where stress factors and their propagation can vary significantly.

Experiments have been conducted to validate CCOSP with alternatives. Results reveal that CCOSP outperforms other algorithms, especially in scenarios with high stress propagation probabilities. In situations where the probability of stress propagation is low, the D-AS converges to No Search, delivering no statistical difference in performance. Hence, D-AS in CCOSP maintains the ability to adapt according to real-time information and the probability of propagating the stress. This finding underscores the potential of CCOSP in optimizing operational efficiency and reducing costs in agricultural monitoring.

The chapter offers contributions to addressing system errors in monitoring environments. The adaptability of CCOSP, capable of adjusting to different stress propagation levels, enhances ARS's performance.

Appendix

Appendix 5.1 Overage Cost (C_O)

Consider a location in a greenhouse. If the ARS inspects a healthy crop plant, the additional cost is equal to

$$C_O = P(\text{Adaptive search is activated.}|\text{Crop plant does not have stress}) \times C_I$$

Moreover, $P(\text{Adaptive search is activated.}|\text{Crop plant does not have stress})$ is, by definition, a type 1 error.

Therefore

$$C_O = \alpha \times C_I$$

■

Appendix 5.2 Underage Cost (C_U)

Consider a location in a greenhouse. If the ARS does not inspect the stress crop plant, the additional cost is equal to

$$C_U = (1 - P(\text{Adaptive search is activated}|\text{Crop plant has stress})) \times C_I$$

Moreover, $1 - P(\text{Adaptive search is activated}|\text{Crop plant has stress})$, by definition, is a type 2 error.

Therefore

$$C_U = \beta \times C_I$$

■

Appendix 5.3 Dynamic Adaptive Search

Consider location m in a greenhouse. The algorithm inspects location $m + 1$ if it expects C_U is higher than C_O, which can be formulated as follows:

$$P(M > m) \times C_U < P(M \leq m) \times C_O$$

In addition, at m^*, the algorithm should stop the inspection, which can be written as follows:

$$P\left(M > m^*\right) \times C_U = P\left(M \leq m^*\right) \times C_O$$

Moreover, as the C_U and C_O are formally formulated, we receive as follow

$$P\left(M \leq m^*\right) \times \alpha \times C_I = P\left(M > m^*\right) \times \beta \times C_I$$

$$\left(M \leq m^*\right) \times \alpha = P\left(M > m^*\right) \times \beta$$

$$P\left(M \leq m^*\right) \times \alpha = \left(1 - P\left(M \leq m^*\right)\right) \times \beta$$

$$F\left(m^*\right) = P\left(M \leq m^*\right) = \frac{\beta}{\alpha + \beta}$$

■

Appendix 5.4 Stress Propagation Model (SPM)

Consider Fig. 5.10 as a model of stress propagation, where p is the probability that a crop plant develops stress; q is the probability that stress propagates to the next location.

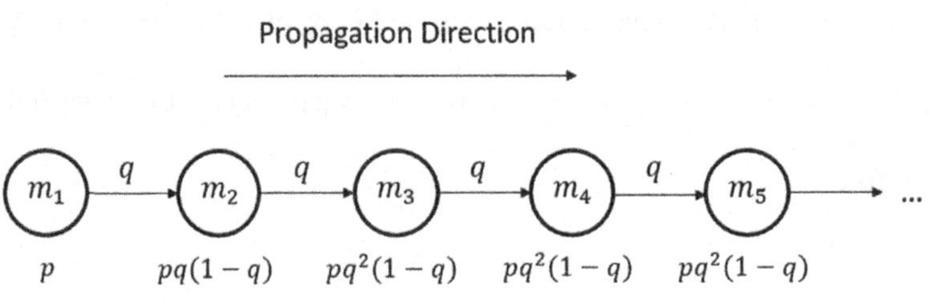

Fig. 5.10 Stress propagation

Table 5.2 Probability that disease propagates to crop plant *m*

$P(M = m)$	Probability that stress propagates to crop plant M
$M = 0$	$(1 - p)$
$M = 1$	$p(1 - q)$
$M = 2$	$pq(1 - q)$
$M = 3$	$pq^2(1 - q)$
...	...
$M = m$	$pq^{m-1}(1 - q)$
...	...

Therefore, the probability that stress propagates to crop plant M, $P(M = m)$, is presented in Table 5.2.

To calculate $P(M \leq m)$, consider:

$$P(M \leq m) = 1 - P(M \geq m + 1)$$

$$= 1 - (pq^m(1 - q) + pq^{m+1}(1 - q) + pq^{m+2}(1 - q) + \ldots)$$

$$= 1 - pq^m(1 - q)(1 + q + q^2 + \ldots)$$

$$= 1 - pq^m(1 - q)(\frac{1}{1 - q})$$

$$= 1 - pq^m$$

Therefore,

$$P(X \leq m) = \begin{cases} 1 - pq^m; \, m \geq 0 \\ 0; \, otherwise \end{cases}$$

■

References

1. Ajidarma P, Nof SY (2021) Collaborative detection and prevention of errors and conflicts in an agricultural robotic system. Studies Inf Control 30(1):19–28
2. Bourland FM, Oosterhuis DM, Tugwell NP (1992) Concept for monitoring the growth and development of cotton plants using main-stem node counts. J Prod Agric 5(4):532–538
3. Chang A, Jung J, Maeda MM, Landivar J (2017) Crop height monitoring with digital imagery from unmanned aerial system (UAS). Comput Electron Agric 141:232–237
4. Claassen R, Cattaneo A, Johansson R (2008) Cost-effective design of agri-environmental payment programs: U.S. experience in theory and practice. Ecol Econ 65(4):737–752

5. Cremer JL, Konstantelos I, Tindemans SH, Strbac G (2019) Data-driven power system operation: exploring the balance between cost and risk. IEEE Trans Power Syst 34(1):791–801
6. Dardonville M, Bockstaller C, Therond O (2021) Review of quantitative evaluations of the resilience, vulnerability, robustness and adaptive capacity of temperate agricultural systems. J Clean Prod 286:125456
7. Dusadeerungsikul PO, Nof SY (2024) Precision agriculture with AI-based responsive monitoring algorithm. Int J Prod Econ, 109204
8. Dusadeerungsikul PO, Liakos V, Morari F, Nof SY, Bechar A (2020) Smart action. In: Agricultural internet of things and decision support for precision smart farming, pp 225–277
9. Dusadeerungsikul PO, Nof SY (2019) A collaborative control protocol for agricultural robot routing with online adaptation. Comput Indus Eng, 135
10. Forstmeier W, Wagenmakers E, Parker TH (2017) Detecting and avoiding likely false-positive findings—a practical guide. Biol Rev 92(4):1941–1968
11. González-Esquiva JM, Oates MJ, García-Mateos G, Moros-Valle B, Molina-Martínez JM, Ruiz-Canales A (2017) Development of a visual monitoring system for water balance estimation of horticultural crops using low cost cameras. Comput Electron Agric 141:15–26
12. Guo P, Dusadeerungsikul PO, Nof SY (2018) Agricultural cyber physical system collaboration for greenhouse stress management. Comput Electron Agric 150:439–454
13. Lysenko V, Opryshko O, Komarchuk D, Pasichnyk N, Zaets N, Dudnyk A (2017) Usage of flying robots for monitoring nitrogen in wheat crops. 2017 9th IEEE International conference on intelligent data acquisition and advanced computing systems: technology and applications (IDAACS), 30–34
14. Mahlein A-K, Steiner U, Hillnhütter C, Dehne H-W, Oerke E-C (2012) Hyperspectral imaging for small-scale analysis of symptoms caused by different sugar beet diseases. Plant Methods 8(1):3
15. Marchand G, Nicot PC, Albajes R, Carisse O (2020) Epidemiology and population dynamics: modelisation, monitoring and management. In: Integrated pest and disease management in greenhouse crops, pp 195–230. Springer International Publishing
16. Nguyen WPV, Dusadeerungsikul PO, Nof SY (2022) Plant stress propagation detection and monitoring with disruption propagation network modelling and Bayesian network inference. Int J Prod Res 60(2):723–741
17. Qin Y, Wang R, Vakharia AJ, Chen Y, Seref MMH (2011) The newsvendor problem: Review and directions for future research. Eur J Oper Res 213(2):361–374
18. Sheriff G (2005) Efficient waste? why farmers over-apply nutrients and the implications for policy design. Rev Agric Econ 27(4):542–557
19. Urruty N, Tailliez-Lefebvre D, Huyghe C (2016) Stability, robustness, vulnerability and resilience of agricultural systems a review. Agron Sustain Dev 36(1):15
20. van Wart J, Kersebaum KC, Peng S, Milner M, Cassman KG (2013) Estimating crop yield potential at regional to national scales. Field Crop Res 143:34–43
21. Wang D, Vinson R, Holmes M, Seibel G, Bechar A, Nof S, Tao Y (2019) Early detection of tomato spotted wilt virus by hyperspectral imaging and outlier removal auxiliary classifier generative adversarial nets (OR-AC-GAN). Sci Rep 9(1):4377
22. Zhang N, Wang M, Wang N (2002) Precision agriculture—a worldwide overview. Comput Electron Agric 36(2–3):113–132
23. Zhang L, Xiao D (2012) Collaborative image compression with error bounds in wireless sensor networks for crop monitoring. Comput Electron Agric 89:1–9

Chapter 6
Cyber-Collaborative Learning Protocol for Precision Agriculture (CCLP)

Abstract This chapter mainly introduces the Cyber-Collaborative Learning Protocol for Precision Agriculture (CCLP). CCLP is the Cyber-Collaborative Protocol (CCP) integrated with Machine Learning and Artificial Intelligence (ML&AI) techniques, enhancing CCP capabilities to indicate directions of stress propagation. Using an agriculture setting as an example, CCLP improves Agricultural Robotic System (ARS) performance by indicating the potential greenhouse stress propagation directions. The stress propagation is modeled as a disruption propagation network, providing a better understanding of the emergence and propagation of crop plant stress. The model provides the foundation for developing an innovative protocol for scanning crop plants. Five different scanning algorithms are developed in this chapter, including one algorithm based on Bayesian network inference analysis, which is used in CCLP as an ML&AI technique, three algorithms derived from disruption propagation network analytics, and a final algorithm that serves as the baseline. The protocol and algorithms have been implemented in agricultural settings, and computer simulation experiments have been conducted. The experiment results indicate that the CCLP outperforms other options and can indicate the potential directions in which stress propagates in a greenhouse.

6.1 Introduction to Cyber-Collaborative Learning Protocols for Precision Agriculture

In the previous chapters, we demonstrated that adopting Cyber-Collaborative Protocols (CCP) has significantly transformed monitoring and regulating crop plants in greenhouses. CCP has effectively enabled synchronization and collaboration among various agents (including humans, robots, and sensors), algorithms, and information flows. This, in turn, has greatly enhanced the performance of Agricultural Robotic Systems (ARS).

In this chapter, we aim to further enhance system performance by leveraging Machine Learning and Artificial Intelligence (ML&AI) techniques, building upon the foundational concepts introduced in earlier chapters. ML&AI will be integrated into

P. O. Dusadeerungsikul and S. Y. Nof, *Cyber-Collaborative Algorithms and Protocols*, Automation, Collaboration, & E-Services 15,
https://doi.org/10.1007/978-3-031-56196-2_6

specific aspects of CCP to elevate system performance. As an illustrative example, we will focus on integrating ML&AI with the Dynamic Adaptive Search algorithm, a component previously discussed in Chap. 5.

One notable application of ML&AI within this framework is its role in guiding the direction(s) of the Dynamic Adaptive Search algorithm, which in Chap. 5 we assume to be known. The status of uninspected crop plants (unknown information) can be inferred (to some degree) from information derived from inspected crop plants or known information [11, 12]. This implementation of ML&AI within CCP leads to the emergence of a Cyber-Collaborative Learning Protocol for Precision Agriculture (CCLP), which will be the central focus of this chapter. Specifically, the Bayesian network inference technique is the ML&AI selected for developing CCLP in this chapter. It is important to note that the development of CCLP presented in this chapter is based on the research conducted by Nguyen et al. [12].

By incorporating CCP and ML&AI, CCLP presents an innovative strategy for addressing the challenges associated with identifying stress propagation direction in crop plants. This chapter offers an exhaustive analysis of CCLP, extending the application of CCP and highlighting how the integration of ML&AI can revolutionize stress propagation identification within greenhouse settings.

6.2 Cyber-Collaborative Learning Protocol in Agricultural Robotic System

6.2.1 Cyber-Collaborative Learning Protocol Framework

We develop the CCLP, which utilizes the capability of ML&AI to guide ARS about the propagation directions of stress in crop plants. The objective of CCLP, in alignment with CCP, is to enhance the efficiency of monitoring systems for crop plants, which can be achieved by integrating cutting-edge agricultural technologies.

One aspect of agricultural automation is using technology to help operations in greenhouses. A robot is equipped with sensors to collect data from crop plants for monitoring their status [4, 6, 10], which are essential in understanding and optimizing the growth and well-being of crop plants.

As previously discussed, CCLP relies on CCP to address concerns related to collaboration and efficiency in identifying stress. To provide a visual representation of this framework, Fig. 6.1 has been included, which is a modification based on the work of Nguyen et al. [12].

Figure 6.1 identifies the essential components of the CCLP. CCLP is responsible for responding to stress propagation by identifying potential inspection directions, which supports the development of effective greenhouse monitoring systems [2, 9, 15, 19].

The integration of ML&AI for indicating the stress propagation directions can significantly improve the ARS capabilities by enhancing the knowledge of stress

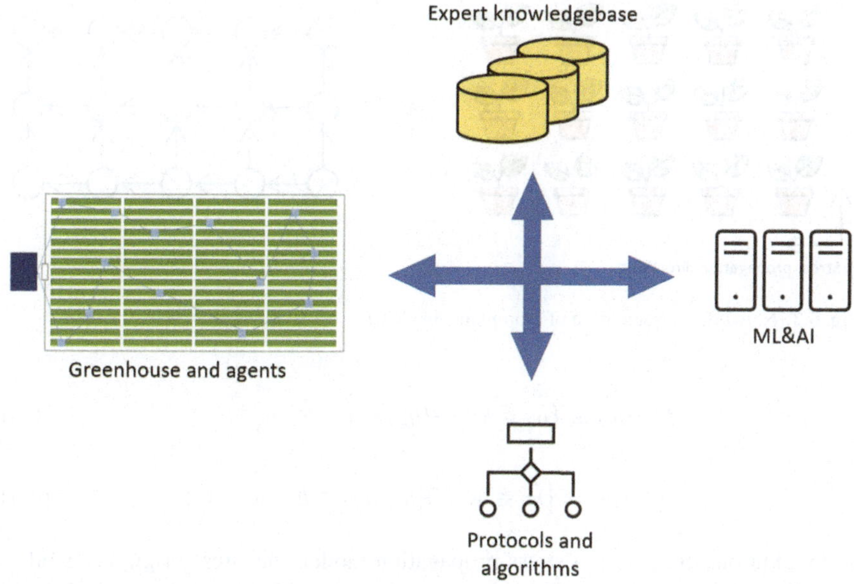

Fig. 6.1 CCLP framework (modified from Nguyen et al. [12])

propagation, which potentially improves strategies and management of crop plants in greenhouses [1, 7, 18].

To identify stress propagation directions, the CCLP involves a set of crop plants organized in a two-dimensional grid denoted by G, with dimensions L and W representing the length and width, respectively. Both L and W are positive integers. Each crop plant is situated within a distinct grid cell and is susceptible to external pressures that may spread in some directions. As a result, the potential stress propagation directions encompass upward, downward, leftward, and rightward. The Collaborative Response to Disruption Propagation (CRDP) modeling framework [13, 14] involves the representation of each crop plant as a node $m \in M$, and the modeling of each potential propagation from one crop plant to another as a directed edge $e = (m_i, m_j) \in E$. Figure 6.2 illustrates the network modeling of crop plants. The illustration involves the conversion of a crop plant grid measuring five by three units, featuring two stress propagation directions, into a network comprising 15 nodes and 22 edges.

Based on the set of nodes (M) and the set of edges (E), Eqs. (6.1)–(6.4) define edges and nodes for each crop plant m.

$$E^{in}(m) = \{(m_i, m_j) \in E : m_j \equiv m\} \tag{6.1}$$

$$E^{out}(m) = \{(m_i, m_j) \in E : m_i \equiv m\} \tag{6.2}$$

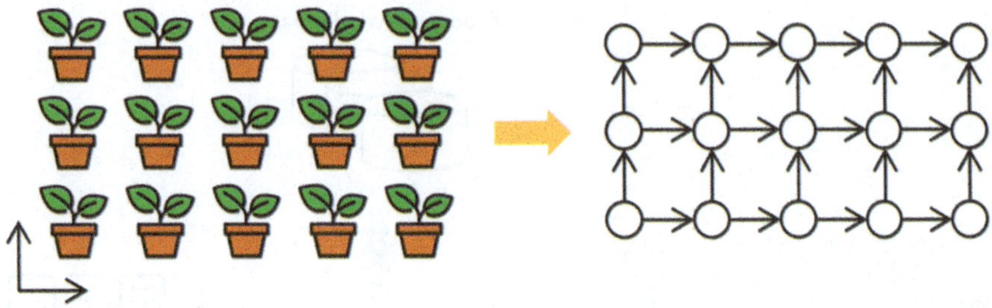

Fig. 6.2 Network representation of crop plants in CCLP

$$M^{in}(m) = \{m_j \in M : \exists(m_i, m_j) \in E : m_j \equiv m\} \tag{6.3}$$

$$M^{out}(m) = \{m_i \in M : \exists(m_i, m_j) \in E : m_i \equiv m\} \tag{6.4}$$

In addition, to develop a stress propagation model, the stress origin probability $p_0 \in [0, 1]$ is defined. This probability represents the likelihood that each stressed crop plant can potentially propagate to neighboring locations. Equation (6.5) defines the origin stress status $S_0(m)$ of crop plant m. Moreover, these stresses at the origin can propagate along the outgoing edges. Each crop plant's final stress status $S(m)$ is defined by Eq. (6.6).

$$S_0(m) = \begin{cases} 1, & \textit{with probability } p_0 \\ 0, & \textit{otherwise} \end{cases} \tag{6.5}$$

$$S(m) = \begin{cases} 1, & \textit{if } S_0(m) = 1 \textit{ or } \exists m_i \in M^{in}(m) : S_0(m_i) = 1 \\ 0, & \textit{otherwise} \end{cases} \tag{6.6}$$

Based on the defined system, Fig. 6.3 shows stress propagation behavior. Initially, three crop plants are affected by $S_0(m) = 1$, then, the stresses propagate, and eight crop plants are affected by $S(n) = 1$.

It is important to note that the value of $S_0(m)$ and $S(m)$ is unknown until a specific crop plant m is inspected. $O(m)$ is the inspection status of crop plant m, and, by

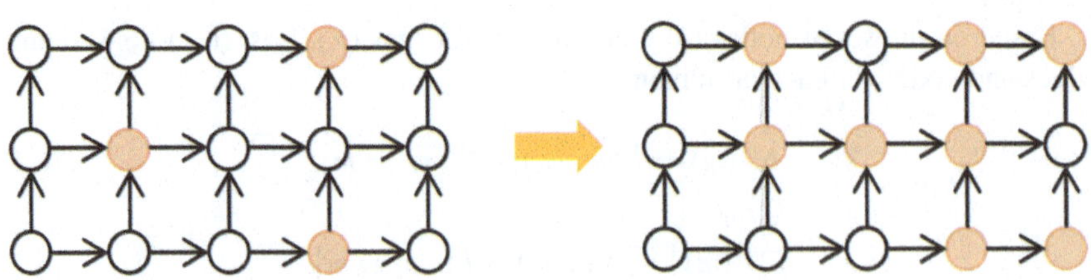

Fig. 6.3 Example of stress propagation

assigning $O(m)$, the status of crop plant m can be identified. Equation (6.7) represents O_m.

$$O(m) = \begin{cases} 1, & plant\ m\ is\ insecpted \\ 0, & otherwise \end{cases} \quad (6.7)$$

Moreover, the number of inspections is usually limited due to the available capacity of agents [5]. Hence, O_b denotes the maximum number of inspections allowed. Equation (6.8) shows the relationship between O_b and $O(m)$.

$$O_b \geq \sum_{m \in M} O(m) \quad (6.8)$$

The following section will discuss the different scanning algorithms for indicating stress propagation direction.

6.2.2 Scanning Algorithm for Indicating Stress Propagation Direction in CCLP

As discussed earlier, the ML&AI is integrated into the CCP to provide stress propagation direction for the Dynamic Adaptive Search algorithm, especially with the multi-directional stress propagation in crop plants. The difficulty of detecting propagation directions consists of determining the scanning assignments based on the current information of stress locations and the underlying network structure [3, 8]. The goal is to utilize information to provide probability to propagate for each direction.

Scanning algorithms have been devised and implemented to address this difficulty. Different algorithms assign scanning duties to agents, implementing different coverage procedures. In addition, different algorithms consider various factors, such as the current stress information of the crop plant, the architecture of the stress propagation network, and other significant parameters, to make informed decisions. These algorithms leverage the integration of the Cyber-Physical System (CPS) and the network modeling capabilities to enhance the overall efficiency and effectiveness of the ARS. Each algorithm is explained in detail in this section.

6.2.2.1 Random Sampling Scanning Algorithm

The first algorithm is the Random Sampling Scanning Algorithm (O_{Random}), serving as the baseline for comparison. O_{Random} involves the location random sampling for performing the inspection duties. The algorithm is straightforward and uncomplicated; therefore, the implementation of O_{Random} may result in a lower level of detection of stress directions, yet easy to implement. It also provides a foundation for evaluating the efficacy of more complex scanning algorithms. O_{Random} can be represented by Eq. (6.9).

$$m_{next} \leftarrow \text{randomly select from } \{m \in M : O(m) = 0\} \tag{6.9}$$

6.2.2.2 Value-Based Scanning Algorithm

The second scanning algorithm is the Value-based Scanning Algorithm (O_{Value}). The algorithm prioritizes crop plants that grow near the identified crop plants to have more chance of being stressed than others. Equation (6.10) (or, equivalently, Eq. (6.11)) presents O_{Value}.

$$m_{next} \leftarrow \underset{m \in M : O(m)=0}{\text{argmax}} \left\{ \left| \left\{ m_i \in M^{in}(m) : O(m_i) = 1 \text{ and } S(m_i) = 1 \right\} \right| \right\} \tag{6.10}$$

$$m_{next} \leftarrow \underset{m \in M : O(m)=0}{\text{argmax}} \left\{ \sum_{m_j \in M^{in}(m)} O(m_i) \times S(m_i) \right\} \tag{6.11}$$

For example, in Fig. 6.4, when the nodes 3, 6, 7, 9, 13, and 15 have been scanned, $O(m) = 1$ for those nodes. After scanning, nodes 3, 6, and 15 are found to be not stressed ($S(m) = 0$), and nodes 7, 9, and 13 are found to be stressed ($S(m) = 1$). Using O_{Value}, m_{next} is determined to be node 8, with a selection value of 2 (due to nodes 7 and 13 being found to be stressed). Other nodes with a selection value of 1 include nodes 2, 4, 10, and 14.

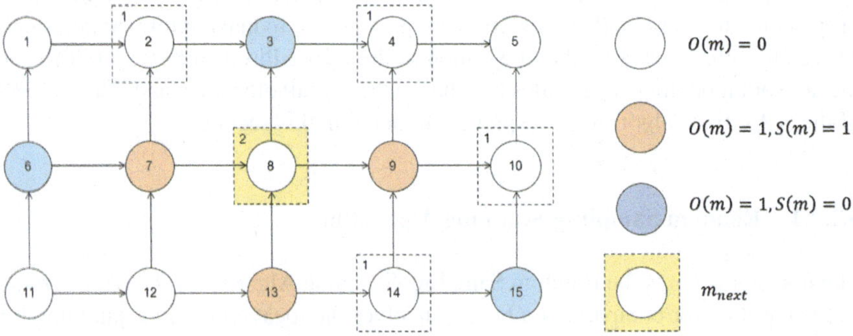

Fig. 6.4 Illustration of value-based scanning algorithm

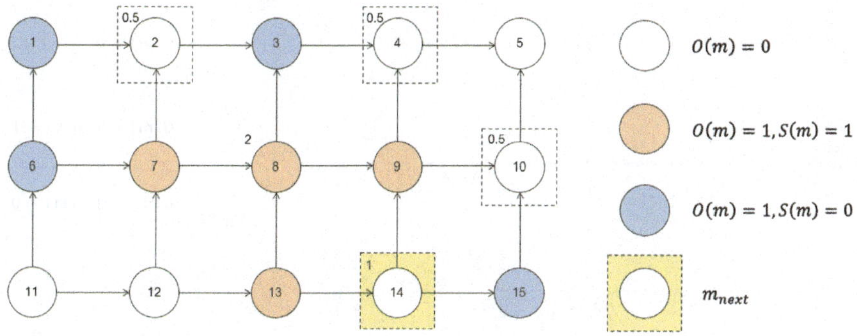

Fig. 6.5 Illustration of Value-based In-degree scanning algorithm

6.2.2.3 Value-Based In-Degree Scanning Algorithm

The third scanning algorithm is the Value-based In-degree Scanning Algorithm ($O_{InDegree}$). $O_{InDegree}$ prioritizes crop plants based on the status of incoming propagation direction. This algorithm expands upon O_{Value} by considering crop plants' status (either indicated stressed or not stressed). w_1 and w_2 are coefficients for adjusting weight between stressed and not stressed crop plants. Equation (6.12) (which is equivalent to Eq. (6.13)) and Fig. 6.5 represents the $O_{InDegree}$.

$$n_{next} \leftarrow \underset{m \in M : O(m)=0}{\operatorname{argmax}} \left\{ \begin{array}{l} w_1 \times \left| \left\{ m_i \in M^{in}(m) : O(m_i) = 1 \text{ and } S(m_i) = 1 \right\} \right| - \\ w_2 \times \left| \left\{ m_i \in M^{in}(m) : O(m_i) = 1 \text{ and } S(m_i) = 0 \right\} \right| \end{array} \right\} \tag{6.12}$$

$$n_{\text{next}} \leftarrow \underset{m \in M : O(m)=0}{\operatorname{argmax}} \left\{ \sum_{m_i \in M^{in}(m)} O(m_i) \times ((w_1 + w_2) \times S(m_i) - w_2) \right\} \tag{6.13}$$

For example, with $w_1 = 1$ and $w_2 = 0.5$, when nodes 1, 3, 6, 7, 8, 9, 13, and 15 have been scanned ($O(m) = 1$). Moreover, the results show that nodes 1, 3, 6, and 15 are found to be not stressed ($S(m) = 0$), and nodes 7, 8, 9, and 13 are found to be stressed ($S(m) = 1$). Using $O_{InDegree}$, m_{next} is determined to be node 14, with the selection value of 1. Other nodes with selection values 0.5 include nodes 2, 4, and 10. Note that, if O_{Value} is used, nodes 2, 4, 10, and 14 would all receive the selection values of 1.

6.2.2.4 Value-Based Dual-Degree Scanning Algorithm

The fourth scanning algorithm is the Value-based Dual-degree Scanning Algorithm ($O_{DualDegree}$). The $O_{DualDegree}$ prioritizes the detection based on the status of both incoming and outgoing propagation directions. This algorithm expands upon

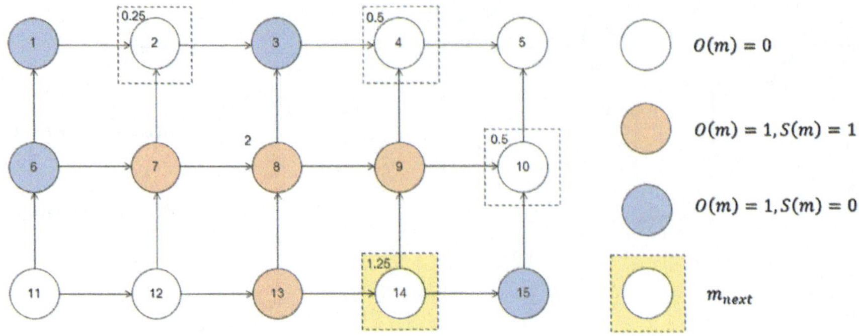

Fig. 6.6 Illustration of Value-based Dual-degree scanning algorithm

$O_{InDegree}$ as it utilizes more information about crop plant status. w_1, w_2, w_3, and w_4 are coefficients for weight adjustment. Equation (6.14), which is also equivalent to Eq. (6.15), presents $O_{DualDegree}$.

$$m_{\text{next}} \leftarrow \underset{m \in M : O(m)=0}{\text{argmax}} \left\{ \begin{array}{l} w_1 \times \left|\{m_i \in M^{in}(m) : O(m_i) = 1 \text{ and } S(m_i) = 1\}\right| - \\ w_2 \times \left|\{m_i \in M^{in}(m) : O(m_i) = 1 \text{ and } S(m_i) = 0\}\right| + \\ w_3 \times \left|\{m_j \in M^{out}(m) : O(m_j) = 1 \text{ and } S(m_j) = 1\}\right| - \\ w_4 \times \left|\{m_j \in M^{out}(m) : O(m_j) = 1 \text{ and } S(m_j) = 0\}\right| \end{array} \right\} \tag{6.14}$$

$$n_{\text{next}} \leftarrow \underset{m \in M : O(m)=0}{\text{argmax}} \left\{ \begin{array}{l} \sum_{m_i \in M^{in}(m)} O(m_i) \times ((w_1 + w_2) \times S(m_i) - w_2) \\ + \sum_{m_j \in M^{out}(m)} O(m_j) \times ((w_3 + w_4) \times S(m_j) - w_4) \end{array} \right\} \tag{6.15}$$

To give an example, with $w_1 = 1$, $w_2 = 0.5$, $w_3 = 0.5$, and $w_4 = 0.25$, Fig. 6.6 shows that m_{next} is node 14 because it received the highest selection value (equal to 1.25), as node 13 was found to be stressed. Moreover, nodes 4, 10, and 2 received 0.5, 0.5, and 0.25 selection values, respectively.

6.2.2.5 Bayesian Network-Driven Scanning Algorithm

The fifth scanning algorithm is the Bayesian Network-Driven Scanning Algorithm ($O_{BayesNet}$), which is considered to be part of ML&AI in CCLP. $O_{BayesNet}$ utilizes Bayesian network modeling [17], inferring the crop plant status of unscanned locations. This algorithm constructs a Bayesian network based on the existing crop plant stress information and uses probabilistic reasoning to estimate the stress status of unexplored areas. By leveraging the ML&AI, i.e., the Bayesian network, the algorithm provides a procedure for predicting the likelihood of stress occurrence in uninspected locations; hence, the propagation directions can be indicated. This information can guide the allocation of scanning tasks and ensure that areas with higher stress probabilities are prioritized. The $O_{BayesNet}$ enhances the detection accuracy and contributes to more efficient resource allocation within the ARS system. O_{BayesNet} is presented as follows.

Bayesian Network-Driven Scanning Algorithm (O_{BayesNet})	
1	Initialize BayesNet
2	$\forall m \in M$, within BayesNet: Create a BayesNet node for $S_0(m)$ Set 2×1 probability table for $S_0(m)$ of BayesNet with $P(S_0(m) = 0) = 1\text{-p}_0$ $P(S_0(m) = 1) = \text{p}_0$ Create a BayesNet node for $S(m)$ Create a BayesNet edge $(S_0(m), S(m))$
3	$\forall e = (m_i, m_j) \in E$, within BayesNet: Create a BayesNet edge $(S_0(m_i), S(m_j))$
4	$\forall m \in M$, within BayesNet: Set conditional probability table for $S(m)$ given all incoming edges: $Number\ of\ rows = 2^{\lvert\{incoming\ BayesNet\ edges\ to\ S(m)\}\rvert}$ $Number\ of\ columns = \lvert\{incoming\ BayesNet\ edges\ to\ S(m)\}\rvert + 2$ For each row of this conditional probability table: If $S(m) = 0$, set $P(S(m) = 0) = 0$ if any incoming BayesNet edge is 1, and 0 otherwise If $S(m) = 0$, set $P(S(m) = 1) = 1$ if any incoming BayesNet edge is 1, and 0 otherwise

In addition, Fig. 6.7 presents an example of 9-node and 12-edge BayesNet. BayesNet begins with initializing the BayesNet (Step 1) to represent the stress propagation of crop plants. Then, in Step 2, a BayesNet node representing $S_0(m)$ for each node $m \in M$ and a corresponding probability table for each node is created. In addition, Step 2 also generates a BayesNet node ($S(m)$) and BayesNet edge (from $S_0(m)$ to $S(m)$). After that, Step 3 refers to the creation of BayesNet edge from $S_0(m_i)$ to $S(m_j)$. By following these 3 steps, a BayesNet can be generated. Finally, in Step 4, the table that contains the conditional probability of all $S(m)$ of all $m \in M$ is created. It is important to note that the dimension of each $S(m)$ depend on the number of incoming BayesNet edge.

Table 6.1 presents a conditional probability table calculated by Step 4 in the O_{BayesNet}, using Fig. 6.4 as an example.

The O_{BayesNet} updates the information about observation ($O(n)$) and $S(n)$ recursively after the inspection at each location has been completed. After that, the probability of being stressed at each uninspected node can be calculated. Note that the belief propagation algorithm [16] can also be applied to calculate the results in polynomial time. The $O_{BayesNet}$ will select a node with the highest probability for the next inspection, given information about locations that were already inspected. Equation (6.16) presents the selection logic of $O_{BayesNet}$.

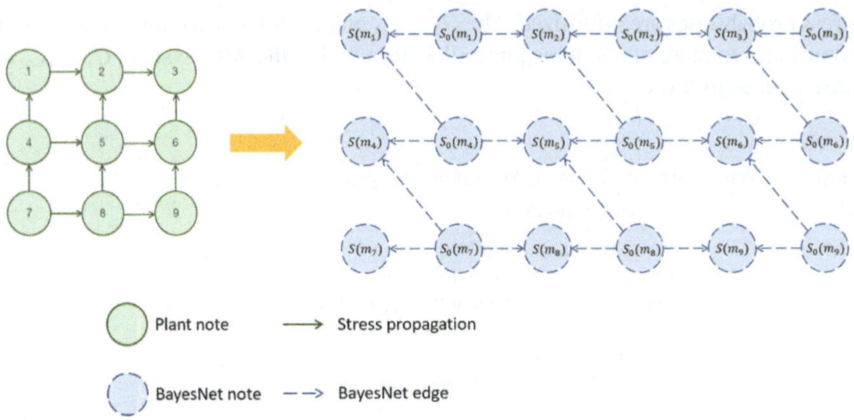

Fig. 6.7 Illustration of Bayesian Network-driven scanning algorithm

$$n_{next} \leftarrow \underset{m \in M : O(m)=0}{\operatorname{argmax}} \left\{ P\left(S(m) | \left\{ S\left(m^*\right) : O\left(m^*\right) = 1, \forall m^* \in N \right\}\right)\right\} \quad (6.16)$$

6.3 Experiment Results and Analysis

ARS with CCLP and other scanning algorithms are validated to evaluate the effectiveness and efficiency. This section provides a summary of the experiment setup and results.

6.3.1 Experiment Settings

The experiments were conducted by computer simulation to validate scanning algorithms. The experimental setting is presented in Table 6.2. Each scenario is subjected to 100 replications, where we randomize the initial stress statuses ($S_0(m)$) for all nodes ($m \in M$). Considering the five stress propagation directions, six cases of p_0, and five scanning algorithms, this comprehensive set of experiments encompasses a total of 150 factorial combinations, resulting in a total of 15,000 simulation runs.

6.3.2 Performance Metrics

Two performance metrics, as follows, are used to evaluate the scanning algorithms in CCLP.

Table 6.1 Results of BayesNet conditional probability from Fig. 6.4

		$S_0(m_{11})$	$P(S(m_{11})=0 \mid S_0(m_{11}))$	$P(S(m_{11})=1 \mid S_0(m_{11}))$
		0	1	0
		1	0	1
	$S_0(m_1)$	$S_0(m_6)$	$P(S(m_1)=0 \mid S_0(m_1), S_0(m_6))$	$P(S(m_1)=1 \mid S_0(m_1), S_0(m_6))$
	0	0	1	0
	0	1	0	1
	1	0	0	1
	1	1	0	1
$S_0(m_4)$	$S_0(m_5)$	$S_0(m_{10})$	$P(S(m_5)=0 \mid S_0(m_4), S_0(m_5), S_0(m_{10})$	$P(S(m_5)=1 \mid S_0(m_4), S_0(m_5), S_0(m_{10})$
0	0	0	1	0
0	0	1	0	1
0	1	0	0	1
0	1	1	0	1
1	0	0	0	1
1	0	1	0	1
1	1	0	0	1
1	1	1	0	1

Table 6.2 Experiment setting

Parameters	Value
Location of crop plants	10×10 grid
Stress propagation directions	4 directions: up, down, left, and right
Search directions	5 directions: 1-direction, 2-direction-opposite, 2-direction-orthogonal, 3-direction, and 4-direction
p_0	0.05, 0.10, 0.15, 0.20, 0.25, 0.30
Scanning budget (O_b)	50
Scanning algorithms	5 scanning algorithms as described in 6.2 are applied: O_{Random}, O_{Value}, $O_{InDegree}$, $O_{DualDegree}$, and $O_{BayesNet}$

6.3.2.1 Undetected Stress Ratio

The Undetected Stress Ratio (*USR*) is the total number of stressed crop plants that were not inspected divided by the total number of crop plants. *USR* can be defined as in Eq. (6.17), equivalent to Eq. (6.18).

$$USR = \frac{|\{n \in M : S(m) = 1 \text{ and } O(m) = 0\}|}{|M|} \tag{6.17}$$

$$USR = \frac{\sum_{m \in M} \max(0, S(m) - O(m))}{|M|} \tag{6.18}$$

6.3.2.2 Redundant Scanning Ratio

The Redundant Scanning Ratio (*RSR*) is the total number of inspected crop plants that were not stressed divided by the maximum number of scans allowed. *RSR* can be defined as in Eq. (6.19), equivalent to Eq. (6.20).

$$RSR = \frac{|\{m \in M : O(m) = 1 \text{ and } S(m) = 0\}|}{O_b} \tag{6.19}$$

$$RSR = \frac{\sum_{m \in M} \max(0, O(m) - S(m))}{O_b} \tag{6.20}$$

Note that both *USR* and *RSR* are minimization objectives. Hence, the scanning algorithm that yields lower *USR* and *RSR* is preferred as it can indicate areas and directions in a greenhouse where crop plants have a higher likelihood to be stressed.

6.3.3 Results and Analysis

In this section, we present the results obtained from the numerical experiments conducted to evaluate the performance of the scanning algorithms. The experimental data collected from the greenhouse environment and scanning algorithms were analyzed to assess the system's effectiveness in detecting stress direction propagation.

6.3.4 Experiment Results

The results shown in Fig. 6.8 represent averages across all scenarios of stress propagation directions and stress origin probability. Notably, the $O_{BayesNet}$ outperforms O_{Random} in terms of the *USR* and *RSR*, both of which exhibit statistical significance. Furthermore, the O_{Value} also shows the improvements over O_{Random}, both in *USR* and *RSR*, and these differences are statistically significant. In contrast, the O_{Value}, $O_{InDegree}$, and $O_{DualDegree}$ do not yield statistically significant performance differences among them.

Lastly, comparing $O_{BayesNet}$ with the next-best scanning algorithms (O_{Value}, $O_{InDegree}$, and $O_{DualDegree}$), the differences in performance are with statistical significance. These findings underline that the $O_{BayesNet}$ not only offers superior stress detection performance, as indicated by the lower number of *USR*, but also demonstrates enhanced scanning efficiency through a reduction in *RSR*.

Moreover, Figs. 6.9 and 6.10 show average values across all stress origin probabilities (p_0) cases, grouped by stress propagation direction. $O_{BayesNet}$ provides the

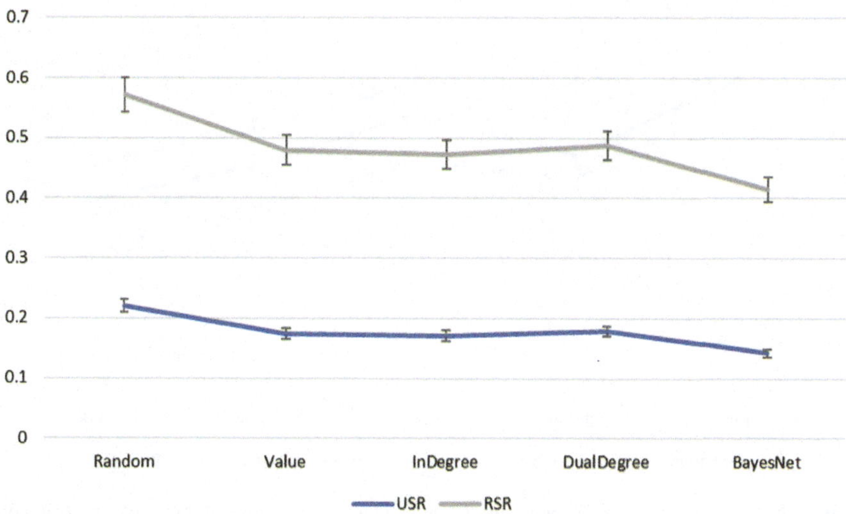

Fig. 6.8 Scanning algorithms comparison

highest performances compared to other algorithms. In addition, the O_{Random} yields the lowest performance in both *USR* and *RSR*.

From Figs. 6.9 and 6.10, when considering the scenario of 1-direction, the next-best scanning algorithm is $O_{DualDegree}$. In the scenario of 2-direction-opposite, the

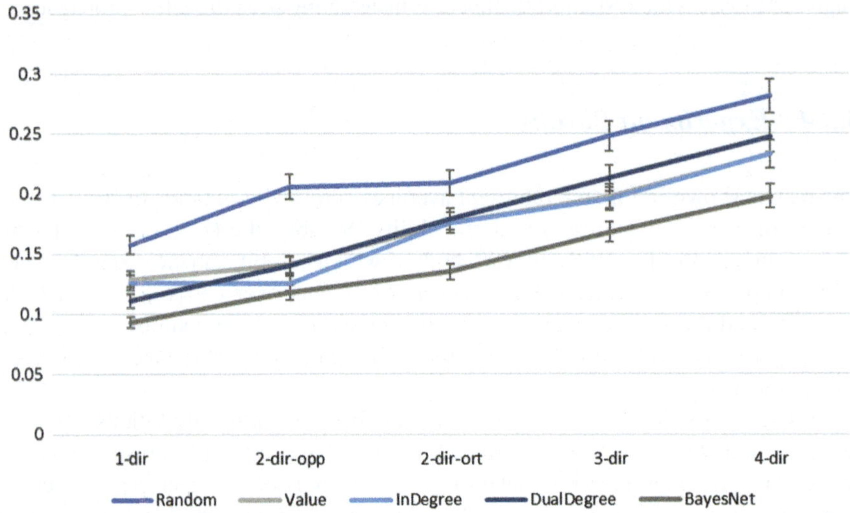

Fig. 6.9 Comparison of ***USR*** between scanning algorithms (grouped by stress propagation directions)

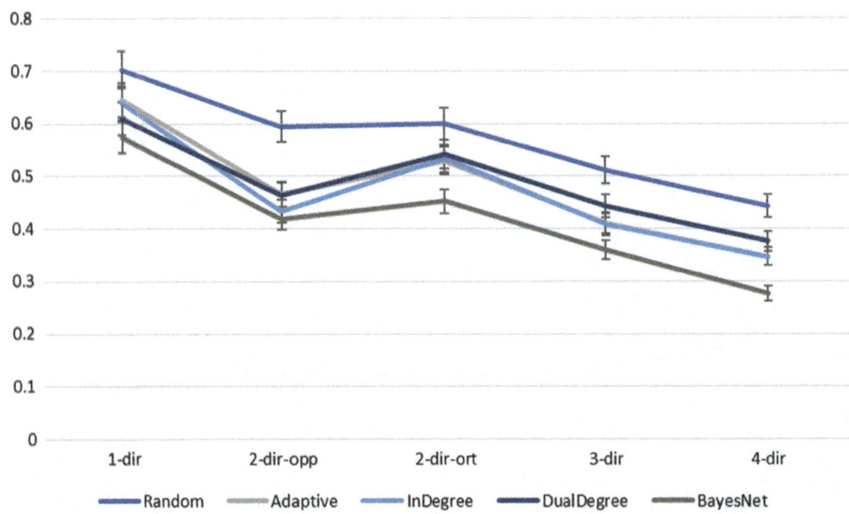

Fig. 6.10 Comparison of ***RSR*** between scanning algorithms (grouped by stress propagation directions)

next-best scanning algorithm is $O_{InDegree}$. In the context of 2-direction-orthogonal, the three scanning algorithms, namely O_{Value}, $O_{InDegree}$, and $O_{DualDegree}$, deliver similar performance. In both *USR* and *RSR*, the performance gap between the two situations of 3-direction and 4-direction, specifically O_{Value} and $O_{InDegree}$, is observed to be minimal, making them the next-best algorithms in terms of performance.

From the experiment results, two observations can be drawn.

1. The $O_{BayesNet}$ demonstrates the highest performance across all scenarios of stress propagation direction.
2. The superiority of the next-best scanning algorithm does not necessarily correspond to the level of advancement (as the second advanced algorithm ($O_{DualDegree}$) does not yield the next-best scanning algorithm in some scenarios).

Figures 6.11 and 6.12 present the performance comparison of scanning algorithms with 95% confidence interval bars. The results are categorized based on the stress origin probability (p_0). In each of the six cases of p_0, the $O_{BayesNet}$ demonstrates superior performance in both *USR* and *RSR*, while the baseline scanning algorithm (O_{Random}) shows the lowest performance, with statistical significance observed in all cases. The differences of *USR* between $O_{BayesNet}$ and the next-best algorithm diminishes as the value of p_0 grows. Furthermore, *USR* exhibits a consistent upward trend as p_0 grows, primarily attributed to the increased in stress locations. In addition, The performance gap between $O_{BayesNet}$ and next-best algorithm in terms of *RSR* also grows as p_0 increases. The *RSR*, however, exhibits a drop trend as p_0 increases, due to the higher number of stresses. The findings indicate that $O_{BayesNet}$ has superior performance compared to alternative scanning algorithms across all p_0 cases. It is noteworthy that the three scanning algorithms, namely O_{Value}, $O_{InDegree}$, and $O_{DualDegree}$, exhibit comparable performance across all p_0.

In summary, the experimental results indicate that the $O_{BayesNet}$ demonstrates superior performance in both *USR* and *RSR* across all experimental scenarios, with statistical significance. Moreover, the O_{Value}, $O_{InDegree}$, and $O_{DualDegree}$ outperform to the O_{Random} in both *USR* and *RSR* in all scenarios, with statistical significance. The statistical significance of the differences in *USR* and *RSR* performance among the three scanning algorithms, namely O_{Value}, $O_{InDegree}$, and $O_{DualDegree}$, is observed only when comparing different stress propagation scenarios.

6.4 Implementation of $O_{BayesNet}$ and CCLP in ARS

The integration and implementation of $O_{BayesNet}$ into the existing CCP is illustrated in Fig. 6.13. The employment of ML&AI into the CCP, yielding CCLP, enhances ARS performance. To achieve seamless operation, a reliable connection between the CCP and ML&AI components is vital.

In this section, we illustrate the interface between $O_{BayesNet}$ (representing the ML&AI part) and the Dynamic Adaptive Search algorithm (representing the CCP

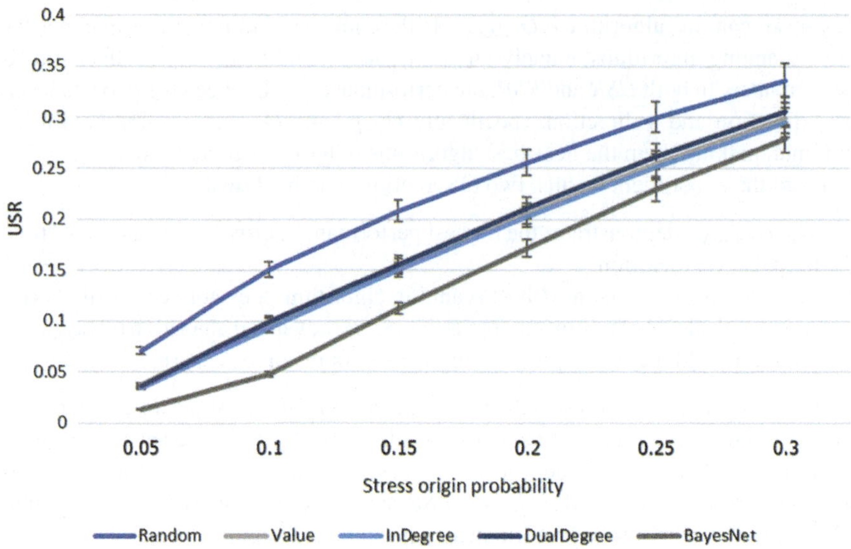

Fig. 6.11 Comparison of ***USR*** between scanning algorithms (grouped by stress origin probability)

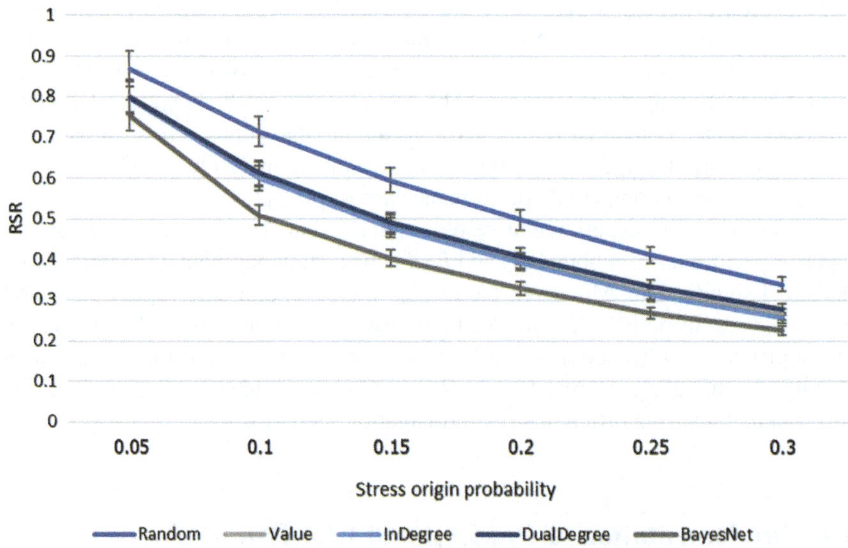

Fig. 6.12 Comparison of ***RSR*** between scanning algorithms (grouped by stress origin probability)

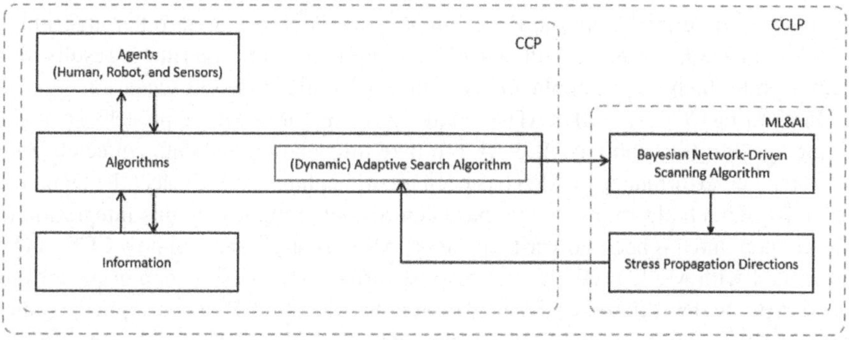

Fig. 6.13 Relationship of $O_{BayesNet}$ in CCLP

part). This linkage is facilitated through the CPS of the ARS and is controlled by established communication protocols within the CCP.

Building on the monitoring system explained in the preceding chapters, the ARS operates through a collaboration among agents, algorithms, and information flow. When the ARS detects stress in a crop plant, the Dynamic Adaptive Search algorithm is triggered. As elaborated in Chap. 5, the Dynamic Adaptive Search algorithm is developed to optimize the trade-off between the number of inspections and associated costs. The direction of the Dynamic Adaptive Search algorithm is guided by $O_{BayesNet}$, which computes the likelihood of stress at unexamined locations.

$O_{BayesNet}$ is a valuable decision support tool, indicating locations within the greenhouse where the likelihood of stress is elevated. By directing the ARS in these specific directions, $O_{BayesNet}$ facilitates more targeted investigations, thereby augmenting the efficiency of the system and minimizing operational time wastage.

6.5 Conclusion and Discussion

The chapter presents an innovative approach in precision agriculture, integrating CCP with ML&AI, resulting in the CCLP. This integration significantly enhances the capabilities of ARS in monitoring and managing stress propagation in crop plants within a greenhouse environment. By leveraging of ML&AI technique, particularly Bayesian network inference, CCLP effectively guides ARS (Dynamic-Adaptive Search algorithm) in predicting potential stress propagation directions in crop plants. Moreover, using a disruption propagation network as a model to represent crop plant stress emergence and propagation offers an innovative approach to a monitoring system. The model has provided the fundamentals for developing a scanning algorithm in the CCLP.

Five scanning algorithms are developed and tested for their efficacy in stress detection and propagation direction indication. Computer simulation experiments are conducted, and the results show the superiority of CCLP, especially because of the algorithm derived from the Bayesian network inference, over other algorithms in terms of *USR* and *RSR*, with statistical significant. The experiment results also demonstrate the practical applications of the CCLP in real-world scenarios.

Integrating CCP with ML&AI techniques has transformed the approach of a monitoring system. By enabling ML&AI to utilize information and data collected from sensors, the performance of ARS has been greatly enhanced. We select the Dynamic Adaptive Search algorithm, which has been impacted mainly by this integration in this chapter and has become a critical part of ARS, as an example of how CCP can be integrated with ML&AI. CCLP can also be further improved by exploring various stress types to develop more advanced protocols. Also, CCLP can be applied to other parts of ARS, such as using ML&AI techniques to develop a protocol for an effective monitoring plan and treatment plan.

References

1. Åstrand B, Baerveldt A-J (2002) An agricultural mobile robot with vision-based perception for mechanical weed control. Auton Robot 13(1):21–35. https://doi.org/10.1023/A:1015674004201
2. Bechar A, Meyer J, Edan Y (2009) An objective function to evaluate performance of human–robot collaboration in target recognition tasks. IEEE Trans Syst Man Cybern, Part C (Applications and Reviews) 39(6):611–620. https://doi.org/10.1109/TSMCC.2009.2020174
3. Chaerle L, Van Der Straeten D (2000) Imaging techniques and the early detection of plant stress. Trends Plant Sci 5(11):495–501. https://doi.org/10.1016/S1360-1385(00)01781-7
4. Diker K, Bausch WC (2003) Radiometric field measurements of maize for estimating soil and plant nitrogen. Biosys Eng 86(4):411–420. https://doi.org/10.1016/j.biosystemseng.2003.08.016
5. Dusadeerungsikul PO, Nof SY (2019) A collaborative control protocol for agricultural robot routing with online adaptation. Comput Ind Eng 135. https://doi.org/10.1016/j.cie.2019.06.037
6. Iida M, Uchida R, Zhu H, Suguri M, Kurita H, Masuda R (2013) Path-following control of a head-feeding combine robot. Eng Agric Environ Food 6(2):61–67. https://doi.org/10.1016/S1881-8366(13)80028-6
7. Keicher R, Seufert H (2000) Automatic guidance for agricultural vehicles in Europe. Comput Electron Agric 25(1–2):169–194. https://doi.org/10.1016/S0168-1699(99)00062-9
8. Mosa KA, Ismail A, Helmy M (2017) Introduction to plant stresses. In: Plant stress tolerance, pp 1–19. https://doi.org/10.1007/978-3-319-59379-1_1
9. Moshou D, Bravo C, Wahlen S, West J, McCartney A, De Baerdemaeker J, Ramon H (2006) Simultaneous identification of plant stresses and diseases in arable crops using proximal optical sensing and self-organising maps. Precision Agric 7(3):149–164. https://doi.org/10.1007/s11119-006-9002-0
10. Nagasaka Y, Zhang Q, Grift TE, Kanetani Y, Umeda N, Kokuryu T (2004) An autonomous field watching-dog robot for information collection in agricultural fields. 2004, Ottawa, Canada August 1–4, 2004. https://doi.org/10.13031/2013.17683
11. Nair AS, Bechar A, Tao Y, Nof SY (2019) The HUB-CI model for telerobotics in greenhouse monitoring. In: Procedia manufacturing, ICPR-25, pp 414–421. https://doi.org/10.1016/j.promfg.2020.01.385
12. Nguyen WPV, Dusadeerungsikul PO, Nof SY (2022) Plant stress propagation detection and monitoring with disruption propagation network modelling and Bayesian network inference. Int J Prod Res 60(2):723–741. https://doi.org/10.1080/00207543.2021.2009139

13. Nguyen WPV, Nof SY (2019) Collaborative response to disruption propagation (CRDP) in cyber-physical systems and complex networks. Decis Support Syst 117:1–13. https://doi.org/10.1016/j.dss.2018.11.005
14. Nguyen WPV, Nof SY (2020) Strategic lines of collaboration in response to disruption propagation (CRDP) through cyber-physical systems. Int J Prod Econ 230:107865. https://doi.org/10.1016/j.ijpe.2020.107865
15. Noon SK, Amjad M, Qureshi MA, Mannan A (2020) Use of deep learning techniques for identification of plant leaf stresses: a review. Sustain Comput: Inform Syst 28:100443. https://doi.org/10.1016/j.suscom.2020.100443
16. Pearl J (2022) Reverend bayes on inference engines: a distributed hierarchical approach. In: Probabilistic and causal inference. ACM, pp 129–138. https://doi.org/10.1145/3501714.3501727
17. Pearl J, Paz A (2022) GRAPHOIDS: graph-based logic for reasoning about relevance relations or when would x tell you more about y if you already know z? In: Probabilistic and causal inference, ACM, pp 189–200. https://doi.org/10.1145/3501714.3501729
18. Reid JF, Zhang Q, Noguchi N, Dickson M (2000) Agricultural automatic guidance research in North America. Comput Electron Agric 25(1–2):155–167. https://doi.org/10.1016/S0168-1699(99)00061-7
19. Wang D, Vinson R, Holmes M, Seibel G, Bechar A, Nof SY, Luo Y, Tao Y (2018) Early tomato spotted wilt virus detection using hyperspectral imaging technique and outlier removal auxiliary classifier generative adversarial nets (OR-AC-GAN). In: 2018 Detroit, Michigan July 29–August 1, 2018. https://doi.org/10.13031/aim.201800660

Chapter 7
Cyber-Collaborative Protocol for System-of-Systems (CCP-SoS)

Abstract This chapter introduces the Cyber-Collaborative Protocol for System-of-Systems (CCP-SoS), which is modified from the protocols developed and implemented by (Dusadeerungsikul et al. [2]). The aim of CCP-SoS is to manage a crop plant treatment team, which is launched after the monitoring process provides the locations where treatment is required. CCP-SoS enhances the crop plant treatment process in greenhouses by utilizing cyber collaboration. CCP-SoS, a Cyber-Collaborative Protocol (CCP) integrated for system-of-systems, has two main parts: CRP-I (Optimization) and CRP-II (Harmonization). CRP-I aims to reduce total operation cost, while CRP-II aims to sequence agents to function smoothly and without conflicts, e.g., conflicting sequence precedence constraints. In this chapter, we first introduce the designed protocol and the related components. Then, we follow with the computer simulation experiments conducted to implement and validate the performance of the designed protocol. The results indicate that the CCP-SoS outperforms alternatives. Because of both CRP-I and CRP-II, CCP-SoS can respond to both planned and unplanned operations. In addition, with human-in-the-loop integrated design, the system can smoothly operate, even if some information is missing.

7.1 Cyber-Collaborative Protocol for System-of-Systems (CCP-SoS)

One of the most important operations in Precision Agriculture (PA) is to apply appropriate farming resources, such as fertilizer, water, and pesticide, at a specific location [12], to treat the required crop plants (e.g., the crop plant with stress). We can consider the following objectives for the treatment team as well as the CCP-SoS framework:

1. Apply fertilizer to specific locations with the lowest overall total operating cost.
2. Minimize total weighted completion time.
3. Minimize makespan.

The first objective follows the main objective of PA and will be achieved by the optimization model. The scheduling of resource applications accomplishes the

P. O. Dusadeerungsikul and S. Y. Nof, *Cyber-Collaborative Algorithms and Protocols*, Automation, Collaboration, & E-Services 15,
https://doi.org/10.1007/978-3-031-56196-2_7

second and third objectives. Because crop plants have different statuses (e.g., different types of stress, different stress levels, etc.), they each have a different priority for specific treatment [14]. Higher priority locations, for example, can face more severe time constraints than regular priority locations (e.g., crop plants with more serious stress or diseases than other crop plants).

Furthermore, during the treatment operation, the system may receive unexpected or emergency requests that often disrupt the current schedule. Using total weighted completion time as one of the system objectives is beneficial because the system will complete the greatest number of tasks (with respect to their priority) during the given scheduled period [9]. As a result of considering the total weighted completion time, if a new request emerges during the execution of a certain treatment process, the new request will have a relatively smaller disruptive effect on the overall schedule.

The following setting will be applied for the rest of this chapter. The CCP-SoS contains two robots, R_1 and R_2, each representing one of two treatments. Hence, it offers three possible robot teaming options: R_1, R_2, and $R_1 + R_2$. Furthermore, there are five task types distinguished based on the ability of either robot or team to accomplish the task.

7.1.1 *Type of Crop Plant Treatment Tasks*

In the treatment application operation process, tasks by robots in a greenhouse can be categorized into five types, explained as follows and in Table 7.1. We follow the task type classification scheme defined by [13] for a collaborative robot team: *Mandatory, Optional, Concurrent, and Concurrent-Mandatory.*

- *Mandatory* type tasks must be executed by a particular robot on the team.
- *Optional* type tasks can be executed by any robot on a given team.
- *Concurrent* type tasks can be executed by alternative sub-teams of robots on the team and/or by the entire team, with varying performances and costs for each sub-team or full-team allocation.
- *Concurrent-Mandatory* type tasks must be executed by a particular, given sub-team, or by the entire team.

Table 7.1 Task types in CCP-SoS

Task type, m	R_1	R_2	$R_1 + R_2$
1	+	+	+
2	+	–	+
3	–	+	+
4	–	–	+
5	+	+	–

This task classification can be expanded to combinations of the four basic types, for instance, task types executed by mandatory sub-teams, optional sub-teams, mandatory entire team, etc.

Similarly, for classifying the crop plant treatment tasks, we define:

Type 1: **Optional**. The first type is the simplest task, which can be executed by either R_1, R_2 or the collaborative team, $(R_1 + R_2)$. This is the most common task type in greenhouses. Examples: applying the common fertilizer that both robots carry.

Type 2: **Robot 1 is mandatory**. In this task type, R_1 must execute the task. The task can either be executed by R_1 or the collaborative team, $(R_1 + R_2)$. Example: applying certain fertilizer that only R_1 possesses and R_2 can serve as its helper.

Type 3: **Robot 2 is mandatory**. Similar to Type 2, R_2 must execute this task type. The task can either be executed by R_2 or the collaborative team, $(R_1 + R_2)$.

Type 4: **Concurrent-Mandatory team type task**. The fourth type of task is the most complicated task, which requires both robots to work together. In other words, the task cannot be executed individually. Example: the location of crop plants that need fertilizers from both robots at the same time.

Type 5: **Optional-Nonconcurrent type task**. In this task type, the task can be executed optionally by R_1 or R_2 individually, but not collaboratively as a team. Example: applying a common fertilizer that both robots carry, where the location has space constraints such as narrow aisles.

7.1.2 Cyber-Collaborative Greenhouse Operations Categories

In CCP-SoS, there are two categories of operations: **Planned operations**, and **Unplanned operations.** In a typical greenhouse, not all operations can be planned in advance, because of the dynamic environment and unexpected, uncertain situations. During the treatment operation as planned according to the monitoring system results, there is a high probability of receiving an additional request, which usually disrupts the process. As a result, the greenhouse should be able to handle unexpected requests, called **unplanned operations**. Each operation category is defined in detail below.

1. **Planned Operations**

Planned operations are included on a list of operations received in advance (from the monitoring process) so the treatment system can plan them before preparing and deploying the treatment team and its resources. Typically, the treatment process begins with receiving the task list, which specifies the form of farming resources, location, and priority. The task list is a list of locations to handle that is normally received ahead of time in accordance with the farming schedule and materials preparation. The task information would define the handling process, indicating the type and number of robot(s) needed. The locations of the given crop plants are the robot's or team's destinations inside the greenhouse. Naturally, the greater the distance between the crop plant locations and the robot-human base point, the higher the operating costs and time [10]. Finally, a given location's priority value indicates the severity of the crop plant status. Relatively higher priorities (higher priority weight) suggest stricter time constraints for this area, and therefore, tasks associated with this location must be completed sooner than others.

2. **Unplanned Operations**

During the treatment process, the system may receive additional requirements called Unplanned Requests (UR). It is important to integrate the UR into the current plan to avoid yield loss. Typically, re-optimizing the entire array of the remaining tasks, including the UR tasks, would yield the best results [6]. Since re-optimization, however, incurs additional costs and requires a relatively long computation time, if the additional costs outweigh the savings from re-optimization, the system should have decision support mechanisms to ensure that an entire schedule of re-optimization is unnecessary. Note that more details about how URs are generated and handled are discussed in Sect. 7.2.

7.1.3 Task Operational Cost (c_{mn})

In this scenario, the Collaboration Requirement Matrix (CRM) specifies the cost of each task to be performed by the robots or their collaborative teams [13]. The example of CRM is shown in (7.1), where c_{mn} is the cost of robots (or team of robots)n performing tasks m.

Due to the collaboration cost for collaborative robot teams, it is assumed that the operation cost of an individual robot to perform a task is strictly lower than the operation cost of the robot team. Hence, when we consider task types 1, 2, and 3, $(m = 1, 2, 3)$ from Table 7.1, where there is an option for assigning an individual robot or a robot team, if a team of robots (meaning that $n = 3$) is assigned to perform the task, the operation cost will be strictly higher than the combination of operation cost performed by the individual robots (due to the additional collaboration cost).

$$CRM = \begin{pmatrix} c_{11} & \cdots & c_{m1} \\ \vdots & \ddots & \vdots \\ c_{1n} & \cdots & c_{mn} \end{pmatrix} \tag{7.1}$$

7.1.4 Task Operational Time (a_{mn})

The task operational time is the total amount of time required to complete the task, which involves receiving the request, preparing the farming resources, traveling to the location, treatment operation, and returning to the base location. If a team of robots performs the task for task types 1, 2, and 3, the operating time will be strictly less than when an individual robot performs the same task (meaning that the team of robots has higher efficiency and better performance). In other words, it is assumed that a team of robots can complete a given task faster than a single robot operation takes to complete the same task [1].

7.2 Cyber-Collaborative Greenhouse System Architecture

Human operators, robots, and sensors are the key agents in the designed CCP-SoS. The system receives feedback from human operators through an interface language, such as VIPO, a spatial-visual programming software [8]. VIPO enables human operators to assign system tasks in a spatial context within an interface, reducing learning time. VIPO generates a computer script specifying the task type, crop plant location, and priority. In addition, VIPO represents the human-in-the-loop (HITL) elements, in which human operators can schedule and assign tasks to robots based on their training, experience, and expertise [19].

If the human operator inputs the task before the operation begins, such task will be a planned operation. On the other hand, if the human operator enters the task after the treatment team is deployed, the task will be classified as UR. All tasks will be entered into the CCP-SoS for managing tasks in real-time, and determining how to optimize overall system output. The main design principle that CCP-SoS utilizes is Collaboration Requirement Planning (CRP). The CRP is divided into two major modules: Optimizer (CRP-I), and Harmonizer (CRP-II) [18]. The optimization module is in charge of assigning tasks to agent, or agents (one-to-one or one-to-many matching) [11]. In other words, the CRP-I is responsible for assigning crop plant locations to the robot or the collaboration robot team. The harmonization module is responsible for sequencing tasks for a robot or team of robots. After the processing of the inputs, CCP-SoS will generate an execution plan. Figure 7.1 depicts the system architecture.

7.3 CCP-SoS Protocol Design

In this section, the CCP-SoS is designed and established. CCP-SoS is the process optimization and coordination protocol for task allocation and task scheduling/sequencing. The CCP-SoS and its components are depicted in Fig. 7.2.

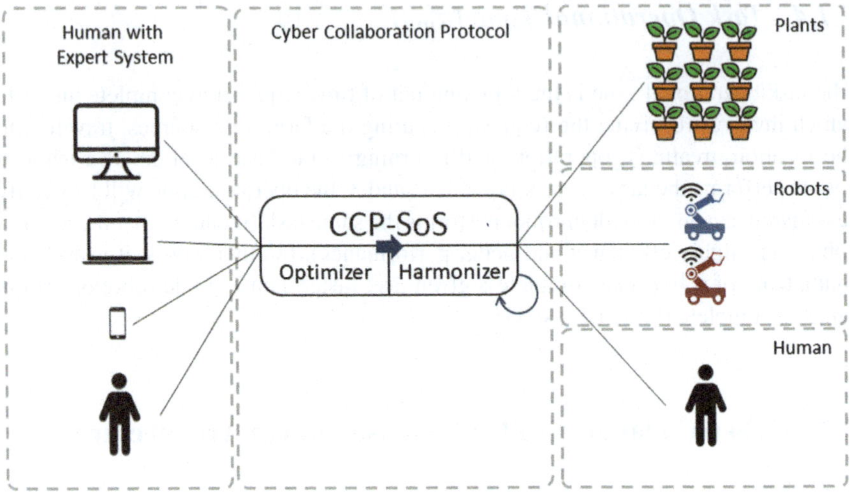

Fig. 7.1 Cyber-Collaborative Greenhouse system architecture

As previously mentioned, CCP-SoS consists of two modules: Optimization (CRP-I), and Harmonization (CRP-II). In the following section, we will elaborate on each component of CCP-SoS.

7.3.1 CRP-I Optimization

The CRP-I aims to keep overall operating costs for planned operations as low as possible. The mathematical model for CRP-I is shown below.

7.3.1.1 Mathematical Model for CRP-I

Let

$$c_{mn} = Cost\ for\ task\ m\ when\ it\ is\ executed\ by\ robot,\ or\ team\ of\ robots,\ number\ n$$

$$x_{mn} = \begin{cases} 1\ if\ task\ m\ is\ executed\ by\ robot,\ or\ team\ of\ robots,\ number\ n; \\ 0\ otherwise \end{cases}$$

$$N_R = Number\ of\ robot\ or\ team\ of\ robots$$

$$N_m = Number\ of\ tasks\ on\ type\ m$$

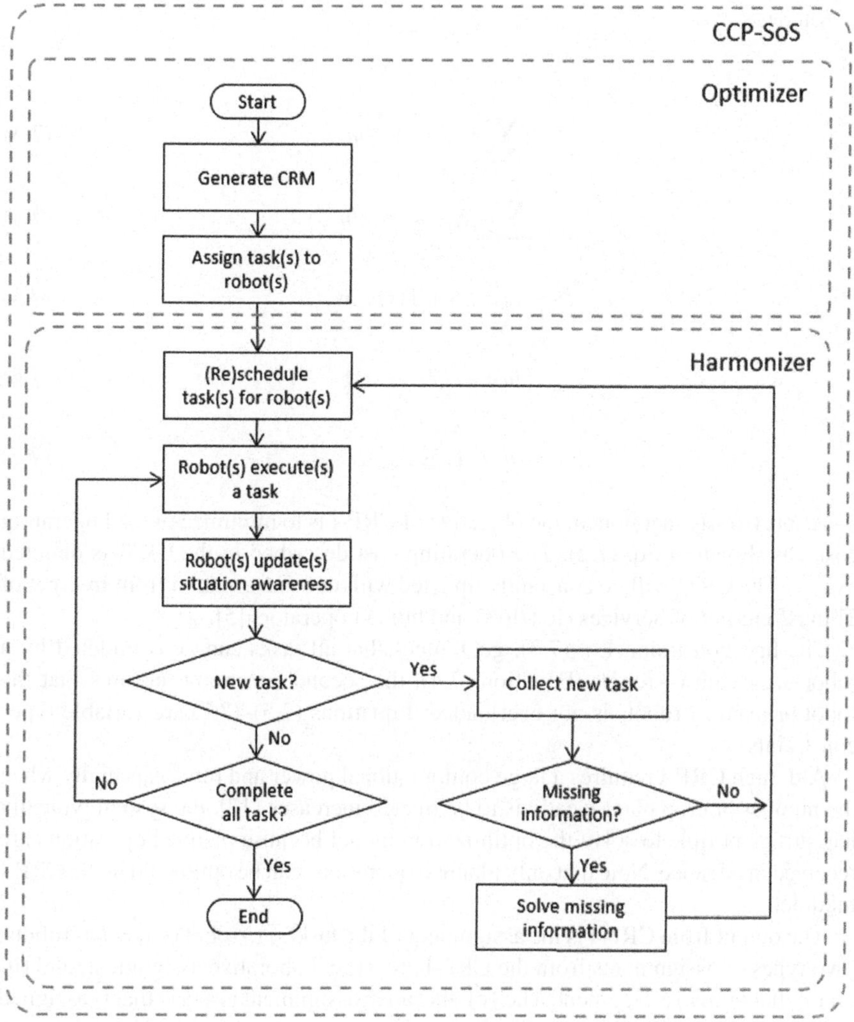

Fig. 7.2 CCP-SoS logic flow diagram

$$N_T = Total\ number\ of\ tasks$$

$$Maximum\ load\ for\ robot\ or\ team\ of\ robots\ = K = \max(\lceil \frac{N_T}{N_R} \rceil, \max_m(N_m))$$

Objective function

$$\min z = \sum_m \sum_n c_{mn} x_{mn} \tag{7.2}$$

Subject to

$$\sum_{n} x_{mn} = 1; \forall m \tag{7.3}$$

$$\sum_{m} x_{mn} \leq K; \forall n \tag{7.4}$$

$$x_{mn} \in \{0, 1\} \forall m, n \tag{7.5}$$

$$m = 1, 2, \ldots, M \tag{7.6}$$

$$n = 1, 2, \ldots, N \tag{7.7}$$

As previously mentioned, the objective of CRP-I is to minimize overall operating costs, as shown in Eq. (7.2). The operating cost described in the CRM is denoted by c_{mn}. The CRM will be constantly updated with new information from Internet of Things/Internet of Services (IoT/IoS) and human operators [5].

The first constraint, Eq. (7.3), guarantees that all tasks can be completed by a robot or a team of robots. Equation (7.4), the second constraint, ensures that the robot or team of robots is not overloaded. Equations (7.5)–(7.7) are variable types constraints.

Although CRP-I requires a large computational power and time, especially when the number of crop plant locations to be treated increases [17], the system typically has sufficient time to solve the optimization model because planned operations are received in advance. Note that only planned operations can be optimized in the CRP-I module.

The output from CRP-I is the assignment of the task(s) to robot or team of robots. Two types of assignments from the CRP-I are: (i) collaborative assignment, and (ii) non-collaborative assignment. The collaborative assignment is a task that is assigned to a team of robot (R_1+R_2), while the non-collaborative assignment is a task which is assigned to a single robot (R_1 or R_2).

The output from CRP-I, however, does not imply the sequence of tasks for each agent. For that result, the harmonization module (CRP-II) is required.

7.3.2 CRP-II Harmonization

Harmonization is the second module of CCP-SoS. The harmonizer's primary objective of the sequencing tasks assigned to each robot, or team of robots, respecting to minimize makespan (C_{max}) and the total amount of weighted completion time

($\sum w_m C_m$). The objective is chosen to ensure that in the event of unexpected circumstances such as UR, robot delays, and operation conflicts and errors, the high priority tasks are completed as soon as possible to avoid heavier losses [16]. In other words, crop plants with relatively more severe status must be treated earlier in the task sequencing.

7.3.2.1 Mathematical Model for CRP-II

Let

$$y = \begin{cases} 1\, if\ task\ m\ starts\ at\ time\ t \\ 0\, otherwise \end{cases}$$

$$p_m = Processing\ time\ of\ task\ m$$

$$w_m = Priority\ of\ task\ m$$

Objective function

$$\min z = \sum\nolimits_m \sum\nolimits_t w_m (t + p_m) y_{mt} \tag{7.8}$$

Subject to

$$\sum_{t=0}^{C_{max}-1} y_{mt} = 1; \forall m \tag{7.9}$$

$$\sum_{m=1}^{M} \sum_{u=\max(t-p_m,0)}^{t-1} y_{mu} = 1; \forall t \tag{7.10}$$

$$y_{mt} \in \{0, 1\}; \forall m, t \tag{7.11}$$

$$m = 1, 2, \ldots, M \tag{7.12}$$

$$t = 0, 1, 2, \ldots, C_{max} - 1 \tag{7.13}$$

The objective function, Eq. (7.8), ensures that the total weighted completion time is minimized. The first constraint, Eq. (7.9), ensures that all tasks begin at the same point in time. Equation (7.10) ensures that only one task can be executed at a time. Equations (7.11)–(7.13) are variable types constraints. The schedule that minimizes the total weighted completion time is created by solving the above mathematical model. A non-delay schedule ensures that the total makespan is kept to a minimum.

It does, however, necessitate a considerable amount of computing power and time, as the mathematical model is NP-hard [15]. As a result, a CRP-II algorithm, called Collaborative Robots Scheduling algorithm (CRS) is developed.

7.3.2.2 Collaborative Robots Scheduling Algorithm (CRS)

Since the harmonization module offers real-time control and adaptation based on new information, it will take place at the local agent(s) level to ensure efficient responsiveness [4]. Solving the previously presented mathematical model (in Sect. 7.3.2.1), which includes a large number of decision variables, necessitates high computational power, which is difficult due to time constraints and local agent limitations. With the goal of harmonization, an algorithm known as the Collaborative Robots Scheduling Algorithm (CRS) is built to ensure that the optimum schedule is accomplished locally at the lowest computational cost.

By developing multi-levels of the Weighted Shortest Processing Time First algorithm (WSPT), the CRS takes advantage, which yields the optimal solution for the total weighted completion time problem. The CRS is defined as follows.

Collaborative Robots Scheduling Algorithm	
1	Initialize
2	Parameter
3	Schedule *collaborative assignment* according to WSPT (this schedule is called collaborative schedule)
4	Schedule *non-collaborative assignment* according to WSPT (this schedule is called non-collaborative schedule)
5	Combine the collaborative schedule with non-collaborative schedule while the releasing time of non-collaborative schedule is the makespan of the collaborative schedule
6	Termination

It can be proven that the CRS algorithm yields the optimal schedule (minimum total weighted completion time). Optimal makespan is ensured by non-delay schedule. In addition, the theoretical guarantee of optimal makespan for collaborative robot schedule is presented in Eq. (7.14).

$$C^*_{max} = \sum_{k \in R_1 + R_2} p_k + \max(\sum_{m = R_1} p_m, \sum_{m\prime = R_2} p_{m\prime}) \tag{7.14}$$

7.4 Human Role and Human-In-The-Loop Design for CCP-SoS

Human support has been an important part of designing, evaluating, and developing production systems [20]. In the CCP-SoS, human is regarded as an intelligent agent [19], capable of having real-time decision-making capabilities [3]. That means a human is able to fill in the missing data, specifying incomplete information. For example, the system can encounter locations that contain incomplete or incorrect priority. In such instances, it is crucial that the human applies their expertise to provide or modify the incorrect or incomplete data, which may include adjusting the location priority, updating the crop plant's location, and adding these locations to the system once the information is completed correctly. Moreover, humans will communicate with the system in real-time, which can help prevent errors and delays in treatment, thus preventing yield loss.

7.5 Experiments and Results

Three computer simulation experiments are conducted to validate the protocol and algorithm developed.

7.5.1 General Experiment Setting

For all experiments, the greenhouse scenario with assigned treatment tasks is simulated. Tasks can be performed in three ways: either by R_1, R_2, or $R_1 + R_2$. The duration of time required at each location is the processing time. The processing time for each task by one of the agents is derived from travel distance plus operation time with the assumption that the collaborative team achieves a faster processing rate than the task executed by either R_1 or R_2.

In addition, as mentioned earlier, different tasks have different operational requirements. The robots' ability to complete a task is assigned to the five types of treatment tasks (Sect. 7.1.1). Weight denoting the relative priority level between zero and ten is assigned to each task. The task with a weight of ten is the highest priority, while the task with a weight of zero denotes the lowest priority.

The cost of operation is directly proportional to time with a fixed cost. Each robot has a fixed cost for every time unit in operation. The cost of the collaboration team is the summation of the participating robots plus the collaborative cost.

Given the experiment settings described above, collaborative tasks incur higher costs in exchange for faster execution. The performance metrics are the overall operation cost, makespan, and weighted completion time. Following CRP, the CRM that represents system status is modified in real-time. As a result, CRM is a dynamic matrix that is administered and updated by the CCP-SoS.

7.5.2 *Experiment 1: Performance Analysis in Planned Operations*

The aim of Experiment 1 is to compare the designed protocol to the common practice procedure (baseline). The overview of the system is as follows.

7.5.2.1 System Description

Agents: 2 robots (R_1; R_2); and 1 robot team ($R_1 + R_2$.)

Tasks: 100 tasks with different priorities and assumed to be ready and available immediately at the beginning of the operation time (all tasks are planned operations).

Operation procedures: Two types of operation procedures.

1. CCP-SoS: The designed protocol.
2. Baseline procedure: Random assignment of task to robot/team of robot, with First Come, First Serve (FCFS) scheduling rule.

7.5.2.2 Results

The outcome of the random task-to-agent assignment with the FCFS rule is used as the design baseline in experiment 1. Performance measurements based on the results of 100 simulation runs are summarized in Fig. 7.3. The average total operation cost and the average weighted completion time of the CCP-SoS are 11.84% and 37.11% lower than the baseline, respectively. At a significance level of 0.05, two sample standard t-test confirm that the CCP-SoS significantly outperforms the baseline in terms of total operating cost and total weighted completion time. Also, CCP-SoS has a statistically significant lower makespan than the baseline. Importantly, the makespan from CCP-SoS in this experiment has reached the theoretical guaranteed optimal makespan from Eq. (7.14).

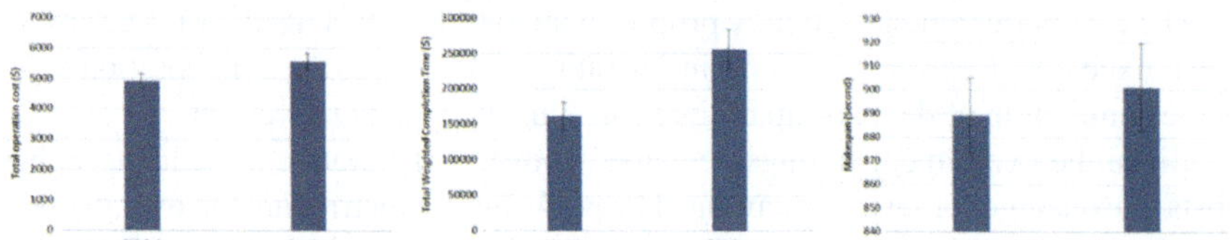

Fig. 7.3 Experiment 1 results

7.5.3 Experiment 2: Performance Analysis with Unplanned Operations

Experiment 2 compares the CCP-SoS to the baseline when unplanned operations occur (i.e., UR). The system description is as follows.

7.5.3.1 System Description

Agents: 2 robots (R_1; R_2), and 1 robot team (R_1+R_2.)

Tasks: 90 tasks are available immediately as planned operations with different priorities, and ten tasks are added after the operation has begun (i.e., ten URs).

Operation procedures: Two types of operation procedures.

1. CCP-SoS: The designed protocol will re-schedule for the new tasks upon arrival, according to CRP-II
2. Baseline procedure: Utilizing CRP-I for the task to robot/team of robot assignment with the First Come, First Serve (FCFS) rule for the new tasks upon arrival.

7.5.3.2 Results

In this experiment, ten tasks (i.e., ten URs) are added to the system after the robot-task assignment and schedule optimization have been completed. The experiment aims to decide the requirements for re-optimization and to understand the effect of dynamic changes in the environment. Figure 7.4 compares performance metrics from 100 simulation runs. Notice that even if only ten tasks are added after the start of the operations and the majority of the task (90 tasks) sequence had been optimized under CRP-I (to investigate the impact of URs and procedures to deal with unplanned operations), the average cost of CCP-SoS is significantly lower than the baseline. Furthermore, the weighted completion time of CCP-SoS is 10.7% lower than the baseline. In practice, the criterion for re-optimization is when the cost savings and weighted completion time outweigh the fixed optimization cost. It is worth noting that the makespan of the CCP-SoS in experiment 2 is greater than the theoretical guaranteed bound from Eq. (7.14). The reason is that Eq. (7.14) captures only planned operations, while the makespan in experiment 2 includes URs, resulting in a larger makespan.

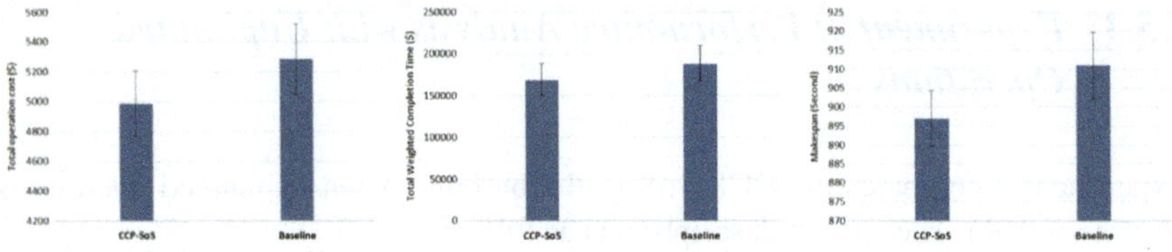

Fig. 7.4 Experiment 2 results

7.5.4 Experiment 3: Performance Analysis with HITL Design

Experiment 3 requires human experts to participate in decision-making. The CCP-SoS, which allows in this experiment for human-in-the-loop features, is validated against the non-human procedure. The overview of the system is as follows.

7.5.4.1 System Description

Agents: 2 robots (R_1; R_2), and 1 robot team (R_1+R_2.)

Tasks: 90 tasks are available immediately, and ten tasks without priority are added after the beginning of the operation (i.e., ten missing priority information).

Operation procedures: Two types of operation procedures.

1. CCP-SoS: The designed protocol with HITL.
2. Baseline procedure: The designed protocol, without HITL.

7.5.4.2 Results

Experiment 3 is designed with ten tasks assigned randomly during the operation process while they have missing information, such as the severity of stress (which represents the task's priority). The experiment investigates the effect of human intervention to supplement the missing information. In the experiments, humans are responsible for inserting missing information based on their expertise. All changes affected by the human agent (which is considered an intelligent agent) are performed in real-time. As a result, the experiment considers the dynamic impact of human intelligence as an expert agent in the loop. Experiment 3 results show that a human agent is capable of leveraging expertise for decisions, which increases the system's robustness compared with a completely autonomous system lacking human inputs. Specifically, the human agent assigns the missing priority of tasks re-scheduling instead of the baseline condition, in which these tasks are inserted at the end of the queue with no priority consideration. Figure 7.5 displays the results of the experiments, indicating that the HITL design substantially reduces overall operating cost and weighted completed time. Since the requested involvement in such cases can be considered part of routine human roles and tasks, it may not incur the additional cost

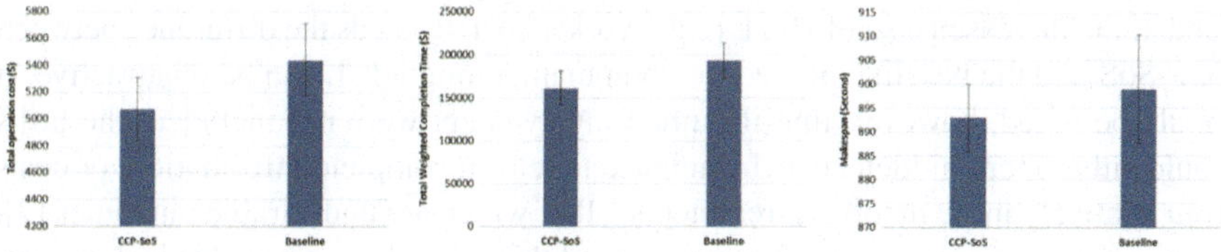

Fig. 7.5 Experiment 3 results

of human involvement. If there is, however, an additional human intervention cost, the additional cost must be less than the difference between the overall operating cost from CCP-SoS and the baseline to be considered a cost-effective situation.

7.6 Conclusion and Discussion

Since stresses are often the first signal of disease, identifying and treating them early is important. To address such a challenge, a new protocol called CCP-SoS is established, developed, and validated, as described in this chapter. The protocol aims at optimizing the treatment team's operation to treat the unhealthy crop plants. Five types of tasks are identified by studying a cyber collaborative system to reflect the greenhouse functional tasks. CCP-SoS comprises two main parts: CRP-I for the task(s) to robot(s) assignment optimization and CRP-II for task sequencing, scheduling, and harmonization. CRP-I operates at the global level of the CCP-SoS, requiring a large amount of computational time, while CRP-II operates at the local-agent level, with only a small amount of computational power to react online to dynamic changes in the greenhouse.

Based on the results of the three experiments, the CCP-SoS outperforms the common practice in terms of total operation cost, makespan, and total weighted completion time. The use of the CRP-I component of the CCP-SoS, which optimally assigns tasks to the robots, enables lower operating costs. Furthermore, the total weighted completion time and makespan are significantly reduced due to the CRP-II component of the CCP-SoS, which can update the execution schedule in real-time through collaborative cyber connectivity with information from IoT/IoS devices. The experiments' results also indicate that HITL design will augment the system to become relatively more robust if the experience and intelligence of human decision-making are used appropriately.

In future practice, the CCP-SoS will save the system both cost and time without requiring any additional investment and, practically, saves the required initial cost investments. Before operations begin, CRP-I can be run in the background. CRP-II, which needs only a small amount of computational power due to its simple algorithm rules, will assist the system by modifying the operation during the ongoing process to respond to URs. Furthermore, with HITL design, the system can effectively solve unforeseen situations, such as missing data, at a low incremental cost. On the other hand, if the increased cost of HITL (e.g., worker cost) exceeds the difference between CCP-SoS and the baseline procedure, then maintaining HITL can be unattractive. It should be noted, however, that if human involvement were not included, the tasks would either incur hidden costs from inaccurate or incomplete information, or costs from relatively more incorrect treatments, all of which are undesirable outcomes [7]. Human agents are an important component of this e-Work or cyber physical system, to help detect erroneous cases and provide them with expert-based information and support, such as a missing priority, thus avoiding further potential losses [19].

References

1. Djuric AM, Urbanic RJ, Rickli JL (2016) A framework for collaborative robot (CoBot) integration in advanced manufacturing systems. SAE Int J Mater Manuf 9(2):457–464
2. Dusadeerungsikul PO, He X, Sreeram M, Nof SY (2022) Multi-agent system optimisation in factories of the future: cyber collaborative warehouse study. Int J Prod Res 60(20):6072–6086. https://doi.org/10.1080/00207543.2021.1979680
3. Dusadeerungsikul PO, Nof SY (2019) A collaborative control protocol for agricultural robot routing with online adaptation. Comput Indus Eng 135.https://doi.org/10.1016/j.cie.2019.06.037
4. Dusadeerungsikul PO, Nof SY (2021) A cyber collaborative protocol for real-time communication and control in human-robot-sensor work. Int J Comput Commun Control 16(3):1–11. https://doi.org/10.15837/ijccc.2021.3.4233
5. Dusadeerungsikul PO, Sreeram M, He X, Nair A, Ramani K, Quinn AJ, Nof SY (2019) Collaboration requirement planning protocol for HUB-CI in factories of the future. Procedia Manuf ICPR-25 39:218–225. https://doi.org/10.1016/j.promfg.2020.01.327
6. Fotio Tiotsop L, Servetti A, Masala E (2020) An integer linear programming model for efficient scheduling of UGV tasks in precision agriculture under human supervision. Comput Oper Res 114:104826. https://doi.org/10.1016/j.cor.2019.104826
7. Gungor ZE, Evans S (2017) Understanding the hidden cost and identifying the root causes of changeover impacts. J Clean Prod 167:1138–1147. https://doi.org/10.1016/j.jclepro.2017.08.055
8. Huang G, Rao PS, Wu M-H, Qian X, Nof SY, Ramani K, Quinn AJ (2020) Vipo. Proceedings of the 2020 CHI conference on human factors in computing systems, 1–13. https://doi.org/10.1145/3313831.3376670
9. Leung JY-T, Li H, Pinedo M (2008) Scheduling orders on either dedicated or flexible machines in parallel to minimize total weighted completion time. Ann Oper Res 159(1):107–123. https://doi.org/10.1007/s10479-007-0270-5
10. Liu J, Li C, Yang F, Wan H, Uzsoy R (2011) Production planning for semiconductor manufacturing via simulation optimization. Proceedings of the 2011 winter simulation conference (WSC), 3612–3622. https://doi.org/10.1109/WSC.2011.6148055
11. Moghaddam M, Nof SY (2017) Mathematical models of best matching, pp 43–62. https://doi.org/10.1007/978-3-319-46070-3_3
12. Mulla DJ (2013) Twenty five years of remote sensing in precision agriculture: key advances and remaining knowledge gaps. Biosys Eng 114(4):358–371. https://doi.org/10.1016/j.biosystemseng.2012.08.009
13. Nof SY, Ceroni J, Jeong W, Moghaddam M (2015) Revolutionizing collaboration through e-Work, e-Business, and e-Service (1st ed.)
14. Oerke E-C, Fröhling P, Steiner U (2011) Thermographic assessment of scab disease on apple leaves. Precision Agric 12(5):699–715. https://doi.org/10.1007/s11119-010-9212-3

15. Pinedo ML (2012) Scheduling: theory, algorithms, and systems. Springer US.https://doi.org/10.1007/978-1-4614-2361-4
16. Potts CN, Van Wassenhove LN (1992) Integrating scheduling with batching and lot-sizing: a review of algorithms and complexity. J Oper Res Soc 43(5):395–406. https://doi.org/10.1057/jors.1992.66
17. Savelsbergh M (1997) A branch-and-price algorithm for the generalized assignment problem. Oper Res 45(6):831–841. https://doi.org/10.1287/opre.45.6.831
18. Scavarda M, Nof SY (2011) Cooperation requirement planning in supply decisions networks—the physical dimension. Proceedings of IIE Annual conference
19. Sreeram M, Nof SY (2021) Human-in-the-loop: role in cyber physical agricultural systems. Int J Comput Commun Control 16(2). https://doi.org/10.15837/ijccc.2021.2.4166
20. Zhang X, Schmidt L, Schlick CM, Reuth R, Luczak H (2008) A human task-oriented simulation study in autonomous production cells. Int J Prod Res 46(18):5013–5041. https://doi.org/10.1080/00207540701222768

Chapter 8
Conclusions, Challenges, and Emerging Trends

Abstract The objective of this chapter is twofold: (1) summarize the concepts and applications of the Cyber-Collaborative Protocol (CCP) as an emerging type of Task Administration Protocol (TAP) to control and optimize complex systems; and (2) summarize the findings and contributions of CCP, using Agricultural Robotics System (ARS) as a representative example of a modern, complex system. In the previous chapters, we present examples of CCP as they contribute to various levels of ARS operation, including system level, agent level, algorithm level, auxiliary level, and system-of-systems level. CCP, utilized as a key tool for planning, optimizing, and harmonizing interactions, shows the contribution and importance of cyber and collaboration in ARS. This chapter also presents the linkage between all the chapters in this book. Lastly, relevant emerging trends are presented so researchers and engineers can prepare for future technological advancements and opportunities. The advancement of new technologies will surely have positive impacts on the design of forthcoming CCP in agriculture, business, industry, services, and government.

8.1 Summary

CCP, derived from TAP with the advancement of Cyber-Physical System (CPS), has become the focus of the recent design of work, factories, installations, and networks, as collaboration and communication between agents and other system components have significantly increased [13]. Complex systems may contain a massive flow of information, multiple agents with decentralized and distributed capabilities, and other advanced technologies [7]. Therefore, CCP is required for systematically managing and optimizing system agents by utilizing the advantage of the cyber connection. Cyber elements, mathematical models, Machine Learning and Artificial Intelligence (ML&AI) techniques, rules, heuristics, and algorithms are essential components of CCP. When the CCP is well-designed, the protocol may be the most effective tool to overcome the multi-scale, multi-level complications arising when collaborating and coordinating complex systems, striving to achieve optimal performance.

P. O. Dusadeerungsikul and S. Y. Nof, *Cyber-Collaborative Algorithms and Protocols*, Automation, Collaboration, & E-Services 15,
https://doi.org/10.1007/978-3-031-56196-2_8

The main objective of CCP is to utilize cyber effectively to manage, both individually and as teams, the proper agents to:

1. Perform the needed tasks correctly at the right time and sequence, and
2. Reduce system waste, such as unnecessary work, excessive cost, and systems Conflicts and Errors (C&E).

CCP has been applied to optimize systems in many different settings (see examples [5, 14, 23, 28, 36]). In this book, we selected Precision Agriculture (PA) as a challenging domain to illustrate the application of CCP at various levels and scales of complexity. PA is an interesting and challenging field because PA encompasses multiple challenges, such as the unstructured and uncertain settings by nature; massive information flows and interactions under time and resource constraints; and multiple agents (including natural resources, human operators, human experts, multiple sensors, robots, and other automation) all working together and in parallel in the systems, with common goals.

Another important aspect of PA is that PA has been chosen and proven as a means to improve food security [20, 21]. Crop plants are grown in greenhouses to protect them as much as possible from undesirable effects. Nevertheless, crop plants are not perfectly controlled and, in many cases, endure stress, which will eventually develop into disease [32]. Timely detection, therefore, is the key to alleviating the damages that would occur without advanced, preemptive mitigation. To apply the recommended (best matched) farming resources for a specific crop plant, PA requires a robotic system called ARS to operate the assigned tasks precisely. Therefore, PA is selected for this book to demonstrate CCP applications in a highly challenging domain, to solve PA issues, and ultimately promote food security—which is one of the current significant global challenges [42].

In this study, we applied CCP in the agricultural setting to support PA at multiple levels and scales and illustrate the benefits of CCP in a complex system. Specifically, the CCP is important in enabling collaborating agents and other components to achieve the system's goals. Table 8.1 briefly highlights the objectives of each chapter. Next, Fig. 8.1 illustrates inter-connections among chapters in this book. More detailed conclusions are presented in the next section.

8.1.1 CCP at System Level

To achieve an effective monitoring system, we applied CCP to the system level by developing agricultural CPS, called CCP-CPS. The main objective of the CCP-CPS is to connect and collaborate cyber and physical agents, knowledge-based information, and other components effectively. Three main physical agents (human, robot, and sensors) are designed in this study to be responsible for different roles. A human operator is the system decision-maker who will solve an unexpected real-time problem [45]. A robot is responsible for carefully moving into a greenhouse and approaching the crop plant's parts for sensors to approach and inspect. Sensors that perform well

Table 8.1 Brief summary of chapters' objectives

Chapters	Chapter's objective
Chapter 1	General background about CCP, Previous research, and applications of CCP in various domains
Chapter 2	The motivation of the selected study, PA, of applying CCP; General description and assumptions of the selected study
Chapter 3	Example of CCP at **system level** (Cyber-Collaborative Protocols for Cyber-Physical Systems: CCP-CPS)
Chapter 4	Example of CCP at **agent level** (Cyber-Collaborative Protocol for Early Detection of Stress in Crop Plants: CCP-ED)
Chapter 5	Example of CCP at **algorithm level** (Cyber-Collaborative Optimal Search Protocols for Precision Agriculture: CCOSP)
Chapter 6	Example of CCP at **auxiliary level** (Cyber-Collaborative Learning Protocol for Precision Agriculture: CCLP)
Chapter 7	Example of CCP at **system-of-systems level** (Cyber-Collaborative Protocol for System-of-systems: CCP-SoS)

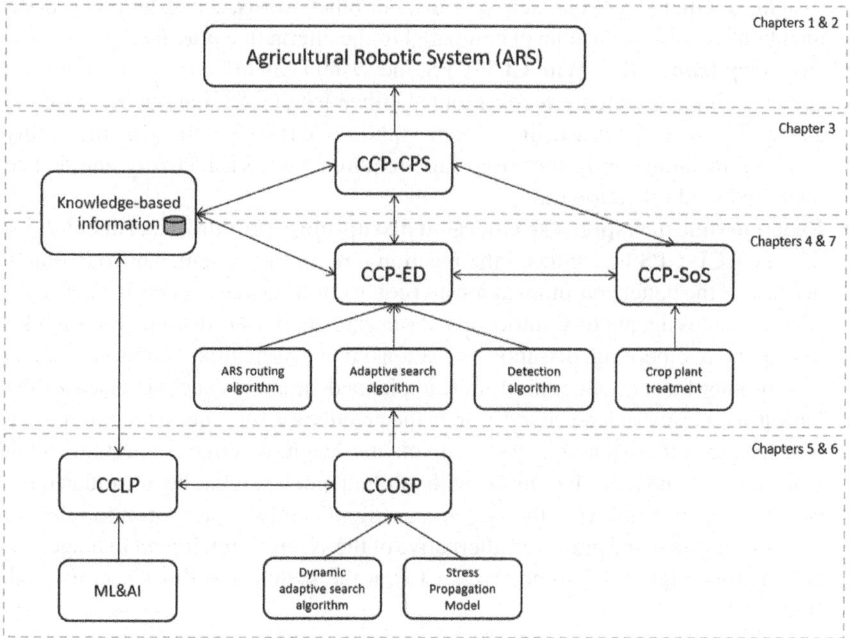

Fig. 8.1 Connection among chapters in the book

in data collection [48] record necessary data, then transmit them to cloud computing, which will be analyzed by the expert system and human agents about the status of crop plants, as well as build and update an online knowledge-based information system. Then, the process repeats until all locations are inspected completely, as assigned.

The CCP-CPS is designed to coordinate collaborating agents, knowledge-based information, and other actuators in agricultural CPS. The experiments are conducted with the following objectives:

- Compare the system's performance with CCP-CPS (Agricultural CPS) employed relative to the current practice (non-CCP and non-CPS).
- **Lower total operation time**: First, results show that Agricultural CPS can yield significantly lower total operation time. The reasons are all associated with an essential capability afforded with CCP: The effectiveness of communications and interactions/connections among main (relevant) agents.
- **Higher tolerance to C&E**: The second experiment shows that a system with CCP-CPS employed has a relatively higher C&E tolerance (error—within a sub-system; conflict—between sub-systems.) In other words, the system can better maintain its performance level compared to the alternative practice.
- **Recovery from C&E**: With CCP-CPS, the system can effectively communicate and coordinate to quickly recover and stabilize when C&E cannot be prevented and do occur. In contrast, the system without CCP-CPS will lack this ability, resulting in significantly increased time required for C&E recovery and, hence, increased total operation time.
- **Reaction time in response to emergent disruptions**: The third experiment shows that the CCP-CPS has reduced the reaction time of the system. This outcome is because of the better communication and connection between agents in CCP-CPS, which enables agents to significantly lower the amount of time the system takes to respond to emerging, disruptive situations. In the agricultural environment, the system may face, at uncertain times, unplanned situations such as diseases that have just emerged at a given location in the greenhouse, or requests for additional, new locations to visit and inspect crop plants. The more effective system should react and respond to such requests with minimum delay. It has been validated that with CCP-CPS employed, the system can significantly improve communication and connection as the preferred alternative of the system's intelligent management and control. Figure 8.2 summarizes CCP at the system level using an example from Chap. 3.

8.1.2 CCP at Agent Level

Although the CCP applied to the system level (CCP-CPS) provides the superiority of the system performance in three perspectives (see Chap. 3), it does not indicate the interaction between local agents and algorithms. To address this objective well,

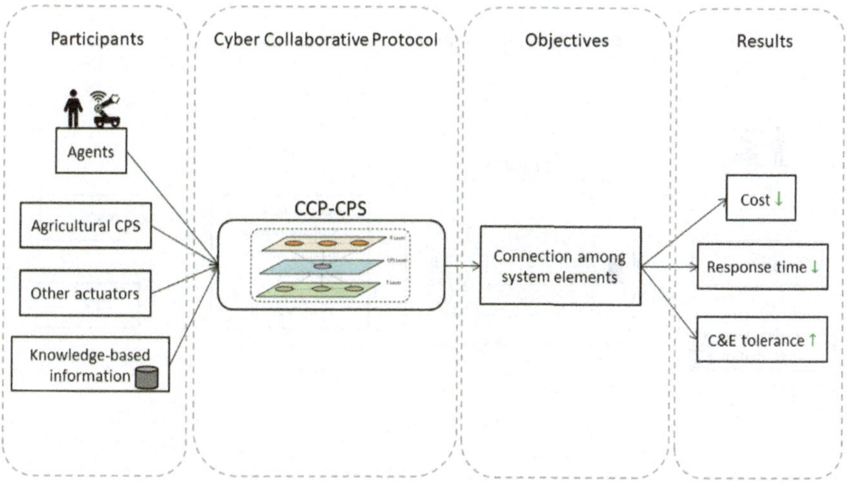

Fig. 8.2 Summary of CCP example at the system level

Chap. 4 illustrates an example of how CCP helps solve the problem at the local agent level.

In Chap. 4, a protocol called CCP-ED is developed to manage agents (human, robot, sensors), algorithms (Routing algorithm, Adaptive Search algorithm, and Detection algorithm), and knowledge-based information to monitor greenhouse crop plants.

Two performance metrics (Number of Infected Crop Plants Found: $NICF$, and Overall Robotic Effectiveness: ORE) have been defined and applied for capturing CCP-ED performance. Computer experiments indicate that the CCP-ED yields better monitoring and inspection performance (represented by $NICF$) and system efficiency (represented by ORE). Because of cyber controlled collaboration among agents and the effective connection to knowledge-based information, the system utilizing CCP-ED can identify and indicate relatively more stressed crop plant locations than the current practice. Moreover, CCP-ED also improves the system's efficiency by minimizing unnecessary work, such as traveling time or data transition time. With that, it provides a system to have more available time to perform more valuable tasks such as inspection, time for or searching for the stressed crop plants. Notably, an example in Chap. 4 illustrates the impact of knowledge-based information on system's performance with the support of CCP-ED. With the linkage between agent and knowledge-based information, the system can maximize scientific knowledge (e.g., higher risk disease propagation directions) to improve the search for stressed crop plants and, as a result, maximize the system's output. Figure 8.3 presents CCP-ED, which is an example of CCP at the agent level.

The CCP-ED provides a guideline for how the agent interacts with other agents, algorithms, and other parts of the monitoring system. An in-depth investigation of the core algorithm would be more meaningful of how CCP can improve system

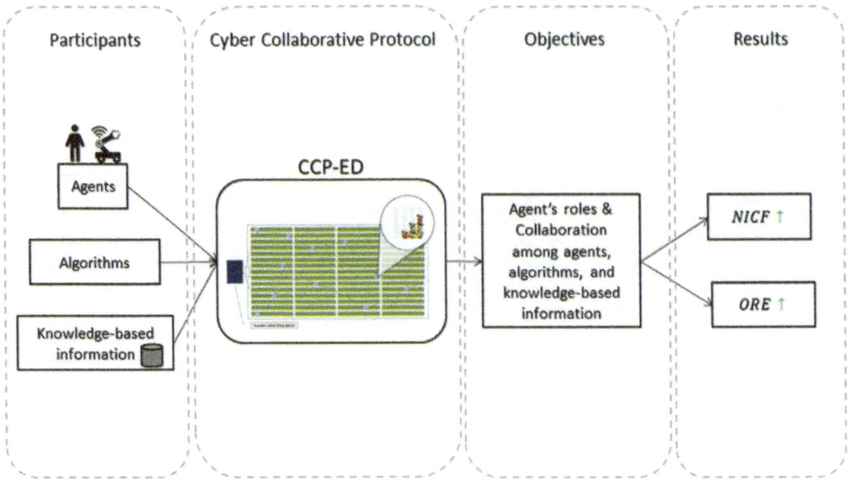

Fig. 8.3 Summary of CCP example in agent level

performance at the algorithm level. In Chap. 5, an Adaptive Search algorithm is selected for further analysis.

8.1.3 CCP at Algorithm Level

The Adaptive Search algorithm (AS) is selected as an illustration for applying CCP because of its complexity: This algorithm contains multiple elements, such as collaboration between robot and sensors, detection algorithm, and knowledge-based information. The objective of the study is to investigate the benefit of CCP at the algorithm level.

The CCOSP, which has two algorithms developed in the study by utilizing CCP, aims to provide an optimal search procedure for minimizing inspection time while maximizing information gain. The CCOSP has two algorithms (1) Dynamic Adaptive Search algorithm (D-AS) and (2) Stress Propagation Model (SPM), connected with CCP. In addition, D-AS incorporates consideration of system error (both type 1 and 2) and stress or disease characteristics (e.g., disease-generated probability and disease-propagation probability) into account.

Three numerical experiments validate the CCOSP. The first experiment focused on performance analysis, aims to investigate the algorithm in terms of the effectiveness of the operation (represented by True Positive Ratio (TPR) and True Negative Ratio (TNR)). The results indicate that the CCOSP delivers significantly better performance compared to the alternatives. The explanation for this advantage is that D-AS in CCOSP adapts itself according to the situation of the disease. For example, when the disease has a high risk (potential) of propagating to the surrounding crop plants, the D-AS will inspect the nearby crop plants relatively more than the situation where the stress is less likely to propagate (according to the most currently available crop plant pathologists expert knowledge).

The second experiment, focusing on cost/banefit analysis, aims to measure the benefit based on the higher amount of useful information gained relative to the additional cost of operation, represented by Monitoring Profit (MP). Different information might yield different benefits to the system. For example, the monitoring system is better to know the locations of crop plants that have stress rather than knowing about the healthy crop plant locations. The reason is that such information gain helps the treatment team plan promptly the precise resources to prevent yield loss (more details in Chap. 7). The experiment results indicate that CCOSP significantly increases MP when compared to the current practice.

Finally, the third experiment, focusing on sensitivity analysis, investigates the performance of CCOSP and alternatives when input deviates from the known value. We let the actual parameter deviate from the setting to observe changes (variations) from the optimal solution in the experiment. The results indicate that CCOSP can deliver the most robust result (less sensitive) compared to alternative search practices. Specifically, when the input parameter deviates 25% from the setting, CCOSP can provide the result within 10% of the optimal solution. This scenario of deviations in parameter settings is common and has a high possibility of happening, especially in the agricultural setting. For example, the probability for the stress to propagate may not be exact as estimated by the input, due to imprecise measurements during movements in the greenhouse. Hence, the less sensitive system is advantageous. In other words, when the actual value has deviated from the setting, a system that can deliver the optimal (or near-optimal) recommendation output is preferred. Figure 8.4 illustrates the conclusion of CCP at the algorithm level.

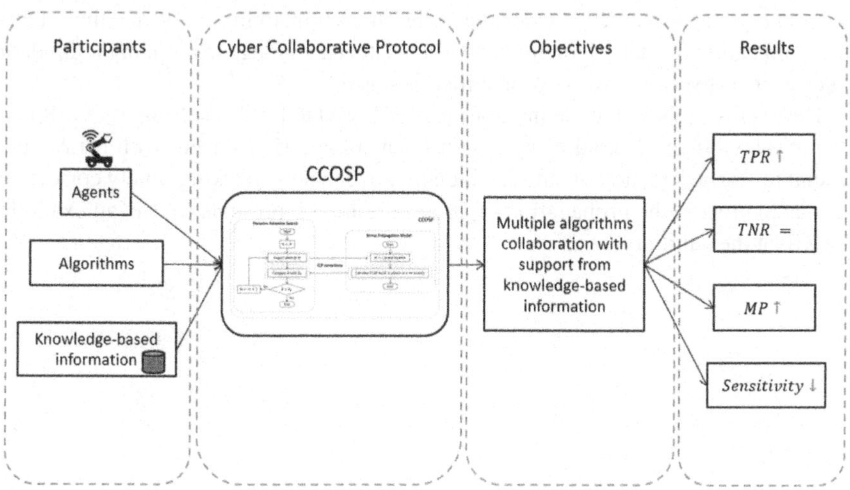

Fig. 8.4 Summary of CCP example in algorithm level

8.1.4 CCP at Auxiliary Level

To improve AS further, Chap. 6 focuses on the development of CCLP to detect stress propagation directions. The stress in crop plants can lead to disruptions that propagate through the network, affecting overall productivity and efficiency.

CCLP is a CCP that integrates ML&AI to utilize system information. As an example, CCLP applies Bayesian network inference as ML&AI technique to detect the stress propagation directions. The CCLP supports CCOSP in the previous development as CCOSP indicates how far the search progression should progress and CCLP indicates the directions of progression. Therefore, CCLP helps ARS to explore only the high potential directions.

Total of five collaborative scanning protocols have been designed and implemented in the experiments. The performance of protocols is evaluated based on two metrics: the number of Undetected Stresses Ratio (USR) and the total number of Redundant Scanning Ratio (RSR), with the goal of minimizing both metrics. The experiment results show as follows.

In terms of overall performance, CCLP is the best-performing protocol, significantly minimizing USR and RSR compared to alternatives. Other alternative protocols which utilize different techniques and information also outperform the baseline protocol but are less effective, compared to CCLP.

When considering performance by different situations of stress propagation directions, CCLP consistently had the highest performance for all cases. Interestingly, the next-best protocol varied depending on the characteristics and direction of the stress propagation scenario, rather than the complexity of the protocol.

In addition, when the initial stress probability varies, the CCLP also outperforms all other protocols. The performance gap (of USR) between CCLP and the next-best protocol, however, decreased as the initial stress probability increased. For RSR, the performance gap increased at when the initial stress probability is low and then stabilized, indicating that CCLP maintained a consistent advantage in reducing redundant scans even as the initial stress probability changed.

The results suggest that the integration of ML&AI to CCP, resulting in CCLP, can lead to more effective monitoring systems. The monitoring system with ML&AI is crucial for early detection of stress, which in turn can help in taking timely corrective actions to prevent the propagation of stress. Figure 8.5 presents CCLP, an example of CCP at the auxiliary level.

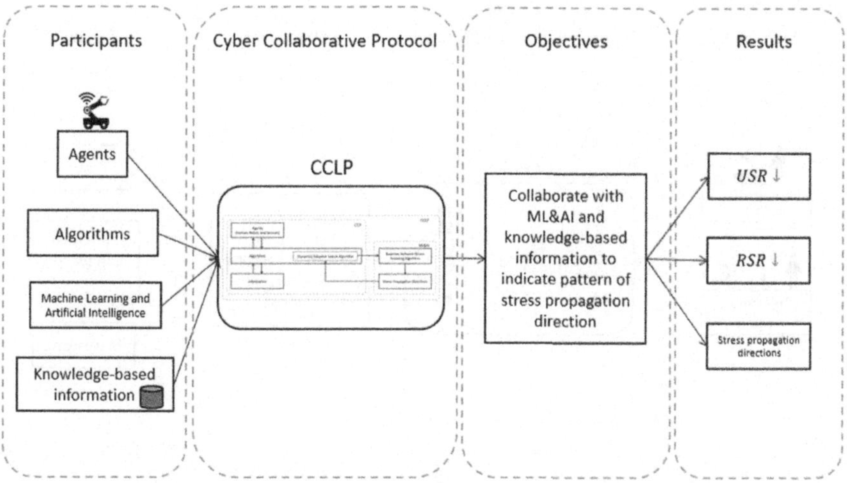

Fig. 8.5 Summary of CCP example in auxiliary level

8.1.5 CCP at System-Of-Systems Level

After the monitoring system indicates the stress or disease locations, another robot team will be launched to treat such crop plants. In this scenario, CCP will manage system-of-systems, connecting the monitoring system with the treatment system. The protocol called CCP-SoS has been developed to show the application of CCP in the system-of-systems perspective. To reflect reality, five task types and two operation categories are defined for representing the greenhouse functional tasks.

CCP-SoS comprises of two main parts: CRP-I and CRP-II. CRP-I is designed to perform at the global level, with high computational power and sufficient computational time. Moreover, it aims to assign tasks to a robot, or robot team, with the objective of minimizing total operation cost.

CRP-II is designed to plan execution and revision under the initial plan that was determined by CRP-I. CRP-II has been designed to perform at a local agent level, because of the benefit of quickly reacting to unexpected real-time issues. Moreover, CRP-II is responsible for sequencing tasks assigned for each robot to minimize weighted completion time and makespan. CRP-II, however, has limited computational power and time, as it performs at the local agent level. Therefore, we develop a heuristic for CRP-II to deliver the recommended solution most promptly.

Experiments are conducted to validate CCP-SoS in different scenarios (e.g., normal operation, operation with urgent requests, and operation with missing information). The system performance is captured with three metrics (1) Total operation cost, (2) Weighted completion time, and (3) Makespan.

The results indicate that CCP-SoS yields superior performance compared to the current practice. CRP-I significantly reduces the total operation cost. In addition, CRP-II, which obtains real-time information and executes scheduling heuristics, can

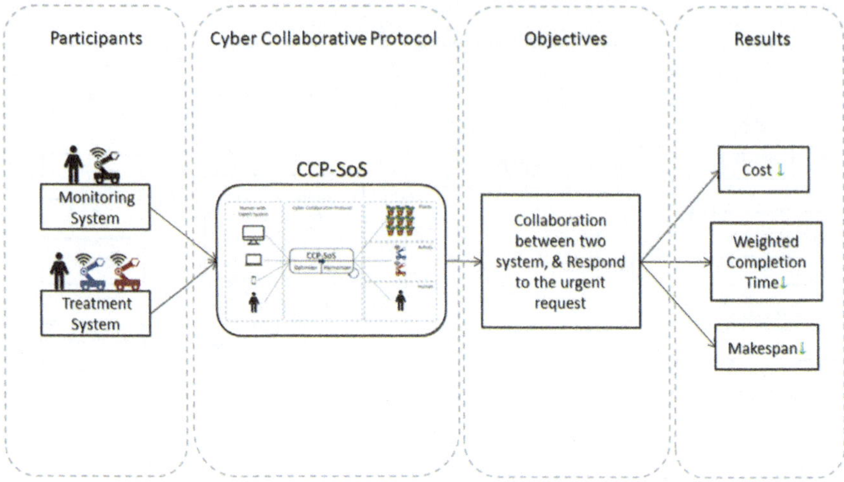

Fig. 8.6 Summary of CCP example in the system-of-systems level

significantly reduce the system's weighted completion time and makespan. CCP-SoS can integrate the new task into the current plan with minimum impact when the system receives an urgent request. Lastly, the human agent plays a significant role in reducing losses that are caused by missing data. Figure 8.6 illustrates the conclusion of CCP at the system-of-systems level.

8.1.6 Key Contributions of CCP

In summary, CCP contributes to improving multiple perspectives of the complex system. With CCP, the system tends to yield better performance because of effective collaboration by the cyber augmentation technology provided through the CCP. The key contributions of CCP to a modern complex system can be summarized as follows.

1. *Effective connection among agents*: CCP enables effective CPS connections and communications among system elements such as agents, knowledge-based information, algorithms, and other actuators. The smooth and reliable connections in CPS enable the system to be more cost-effective [34] as connections and/or communications costs are usually lower than a rework cost [22]. In addition, because the communication cost in CPS is relatively low compared to total operation cost [48], the system can adjust the process when C&Es emerge to avoid excessive costs [16].
2. *Synchronization and harmonization of agents*: When agents do not work collaboratively in a synchronized and harmonized way, a multi-agent system can deliver sub-optimal outcomes [8]. CCP is a crucial tool to synchronize and harmonize agents, knowledge-based, and other system elements to work together [28]. In

addition, in many cases, the collaboration of agents by CCP yields a better system performance than when each agent works individually [15, 17]. Importantly, a system that comprises of multiple weak agents (rather than one powerful agent) usually has a better cost-effective outcome [36] and better system resilience [40].

3. *Information sharing in decentralized and distributed systems*: Because of the cyber connections among agents, algorithms, and knowledge-based information utilizing CCP, agents are able to obtain and learn data from their peers [37]. Such ability strengthens the system as it does not rely only on a single agent. The copies of data are distributed across the system agents, developing a more robust system. This promotes the distributed and decentralized ability of the system, enhancing system flexibility and stability [29].
4. *Real-time corrections*: Even though we are progressing to automation 5.0, C&Es are unavoidable in any system, especially in a system with multiple agents [2]. The system should prepare for C&Es, and, importantly, when C&Es emerge, the system should prevent and resolve them in a real-time fashion [15]. Because CCP takes advantage of cyber to manage the system's components, the C&Es of the system are minimized, either by preventing them, or by effective and timely repair and recovery. In addition, the system will also have a prompt response to identify and re-operate a necessary task; hence, the system has a real-time correction ability when needed.

8.2 Emerging Trends

This book has presented the new emerging tool, CCP, for design and application of augmented collaboration with the advancement of new technologies. PA is the selected domain of illustration to represent a complex system, and how it can significantly benefit by CCP in multiple dimensions. To this end, the following emerging trends can potentially impact further development and scales of CCP.

8.2.1 Big Data, Machine Learning and Artificial Intelligence

The first emerging trend to be considered includes big data as well as Machine Learning and Artificial Intelligence (ML&AI). The complex modern system has a high potential to change over time because of the dynamic environment and interactions [19, 44]. ML&AI can improve the CCP further by enabling the CCP's components, e.g., heuristics, rules, and agents, to evolve when the system obtains more experience and information [26]. Because of the current technologies, massive data can be collected, stored, and, most importantly, utilized [38]. CCP can utilize such advantages to improve the decisions recommended for the system. ML&AI are already emerging parts of the CCP algorithms, protocols, and interfaces, e.g., the preemptive learning protocol developed by [1, 4], and [33]. ML&AI (with support from big data) are viewed as critical directions to deal with dynamic and uncertain changes in the environment.

In addition, CCP with ML&AI integration can assist the main agents when there are missing, unavailable information and logic [12, 46]. Let us use agriculture as an example. In many cases, a human agent may not be available at the time to make the currently necessary decision or information augmentation. ML&AI can suggest the appropriate action to other agents in order to solve the real-time problem [6]. ML&AI, therefore, will be able to create more flexibility and robustness in the system [10]. Such ability will maintain system performance, even though system components have deviated from the initially designed plan.

8.2.2 Factories of the Future and Automation 5.0

The second emerging trend is Factories of the Future (FoF) and Automation 5.0. They are the advancements in technologies that can improve CCP in terms of agent capabilities. The rapid advancement of cyber and digital technology, which leads to Automation 5.0 [35], has brought a new challenge to researchers and engineers [15]. Cyber and digital technology advancements have accelerated the design and implementation of FoF and, consequently, CCP. Consider, for example, the development of Collaborative Intelligence (CI) and its application in decentralized and distributed complex systems, e.g., the design of:

1. **Hub-to-hub, HUB-CIs, and HUBs of CI** [18] where each HUB-CI integrates intelligence flow utilizing CCPs in the system-of-systems, as well as with other HUB-CIs of other networked systems of systems. An example in the scope of this book can be intelligence flows among multiple ARSs in several greenhouses. There can be significant advantages by the HUB-CI to HUB-CI networking, as validated in feedforward alerts for rescheduling when delays occur at earlier stages of a production and supply multi-facility chain (see examples in [39, 41]). The emerging technology of digital twins as emulators of physical system elements can interact with the HUB-CIs to exchange real-time intelligence.
2. **Augmented Reality (AR) and Virtual Reality (VR)**, systems that are already applied for information exchanged with workers in FoF [11]. Increasingly, they are combined with decision support and knowledge-based information to further augment the collaboration of human workers with robots and machines in FoF (e.g., [3, 9, 35, 46]). CCPs are already being designed to synchronize and harmonize the real-time allocation of specific AR to individual workers based on their current tasks and customized to their own skills and abilities. (e.g., [25]), and CCP with HUB-CI augmentation [31].

Inevitably, FoF will have multiple agents which are distributed, decentralized, and network-connected. In addition, massive information flow and interaction between agents are expected. CCP, therefore, will play a key role in managing and optimizing agent, interaction, information, and intelligence flows in FoF.

FoF and Automation 5.0 also impact the design and implementation of CCP. Although CCP discussed in this book is already utilizing the developments of Automation 5.0, future advancements in cyber and digital technology will further enhance CCP's ability. For example, advancement in Automation 5.0 with quantum communications may enable faster information transferred for better decisions in ARS, enabling a human agent to control multiple greenhouses (and multiple ARS) simultaneously.

8.2.3 Blockchain Technology

Blockchain is one of the most noticeable emerging technologies in recent years. Blockchain is defined as a public, permanent, append-only disrupted ledger [30]. As a result, blockchain is well-known for financial technology. Even though most people know blockchain from Bitcoin and other cryptocurrencies, the implication of blockchain technology is significantly broader. Application areas such as smart contracts; supply chain negotiations, agreements, and management; and gaming will soon emerge [24].

Blockchain technology can impact and improve the design and development of CCP. The essential aspects of blockchain technology are decentralization, immutability, security, and transparency [27], which will strengthen the CCP's ability. First, both CCP and blockchain contain and follow decentralization principles and aim not to let any single agent control all activities. Likewise, blockchain technology will significantly improve the decentralized ability of CCP, especially for the decentralized agent and multi-HUB-CIs networks. That is, blockchain (with ML&AI) enables intelligent decentralized autonomous agents [43], which can quickly and automatically validate data in the CCP. In addition, critical and sensitive data in CCP can ensure immutable ability and increase their security by utilizing blockchain technology. The data will not be stored in the central database but at the local node (such as system agents). Hence, even when some agents fail or break down, copies of the data can be retrieved from other agents. Lastly, all data, information, actions, and interactions in the CCP can be systematically traced back because all activities are recorded and stored as blocks and linked with one another in the blockchain, increasing the system's transparency.

We can see that emerging trends and technologies can improve CCP from several perspectives. While applying CCP to improve and even optimize the system flows, intelligence flows, and interactions, engineers and researchers are able to utilize the advantages of new technology for better design, efficiency, effectiveness, agility, and resilience of systems and networks.

References

1. Ajidarma P (2017) Multi-sensor fault tolerant learning algorithm in an agricultural robotic system [M.S. Thesis]. Purdue University Graduate School
2. Ajidarma P, Nof SY (2021) Collaborative detection and prevention of errors and conflicts in an agricultural robotic system. Studies Inf Control 30(1):19–28
3. Ajidarma P, Nof SY (2023) Skill and knowledge sharing as collaborative intelligence in future work and factories. In: Huang CY, Yoon S (Eds), Systems collaboration and integration: past and future research from the PRISM Center
4. Ajidarma P, Nof SY, Pradana RA, Nugroho WA, Halim AH (2022) Preemptive demand and capacity sharing learning protocols using long short-term memory (LSTM) neural network Autoencoders. IFAC-PapersOnLine 55(10):1798–1803
5. Anussornnitisarn P, Nof SY, Etzion O (2005) Decentralized control of cooperative and autonomous agents for solving the distributed resource allocation problem. Int J Prod Econ 98(2):114–128
6. Boutaba R, Salahuddin MA, Limam N, Ayoubi S, Shahriar N, Estrada-Solano F, Caicedo OM (2018) A comprehensive survey on machine learning for networking: evolution, applications and research opportunities. J Internet Serv Appl 9(1):16
7. Cao Y, Yu W, Ren W, Chen G (2013) An overview of recent progress in the study of distributed multi-agent coordination. IEEE Trans Industr Inf 9(1):427–438
8. Cardon A, Galinho T, Vacher J-P (2000) Genetic algorithms using multi-objectives in a multi-agent system. Robot Auton Syst 33(2–3):179–190
9. Chidambaram S, Huang H, He F, Qian X, Villanueva AM, Redick TS, Stuerzlinger W, Ramani K (2021) ProcessAR: an augmented reality-based tool to create in-situ procedural 2D/3D AR Instructions. Designing Interactive Syst Conf 2021:234–249
10. Cioffi R, Travaglioni M, Piscitelli G, Petrillo A, De Felice F (2020) Artificial intelligence and machine learning applications in smart production: progress, trends, and directions. Sustainability 12(2):492
11. de Souza Cardoso LF, Mariano FCMQ, Zorzal ER (2020) A survey of industrial augmented reality. Comput Ind Eng 139:106159
12. Duan Y, Edwards JS, Dwivedi YK (2019) Artificial intelligence for decision making in the era of Big Data—evolution, challenges and research agenda. Int J Inf Manage 48:63–71
13. Dusadeerungsikul P (2020) Operations analytics and optimization for unstructured systems: Cyber collaborative algorithms and protocols for agricultural systems [PhD Dissertation]. Purdue University
14. Dusadeerungsikul PO, Nof SY (2019) A collaborative control protocol for agricultural robot routing with online adaptation. Comput Indus Eng, 135
15. Dusadeerungsikul PO, Nof SY (2021) A cyber collaborative protocol for real-time communication and control in human-robot-sensor work. Int J Comput Commun Control 16(3):1–11
16. Dusadeerungsikul PO, Nof SY (2024). Precision agriculture with AI-based responsive monitoring algorithm. Int J Prod Econ 109204
17. Dusadeerungsikul PO, Nof SY, Bechar A, Tao Y (2019) Collaborative control protocol for agricultural cyber-physical system. Procedia Manuf, ICPR-25, 39:235–242
18. Dusadeerungsikul PO, Sreeram M, He X, Nair A, Ramani K, Quinn AJ, Nof SY (2019) Collaboration requirement planning protocol for HUB-CI in factories of the future. procedia manufacturing, ICPR-25 39:218–225
19. Enke J, Glass R, Kreß A, Hambach J, Tisch M, Metternich J (2018) Industrie 4.0—Competencies for a modern production system. Procedia Manuf 23:267–272
20. Erickson B, Fausti SW (2021) The role of precision agriculture in food security. Agron J 113(6):4455–4462
21. Gebbers R, Adamchuk VI (2010) Precision agriculture and food security. Science 327(5967):828–831
22. Gopal A, Mukhopadhyay T, Krishnan MS (2002) The role of software processes and communication in offshore software development. Commun ACM 45(4):193–200

23. Guo P, Dusadeerungsikul PO, Nof SY (2018) Agricultural cyber physical system collaboration for greenhouse stress management. Comput Electron Agric 150:439–454
24. Hewa T, Ylianttila M, Liyanage M (2021) Survey on blockchain based smart contracts: Applications, opportunities and challenges. J Netw Comput Appl 177:102857
25. Huang G, Rao PS, Wu M-H, Qian X, Nof SY, Ramani K, Quinn AJ (2020) Vipo: spatial-visual programming with functions for robot-IoT workflows. Proceedings of the 2020 CHI conference on human factors in computing systems, 1–13
26. Jordan MI, Mitchell TM (2015) Machine learning: trends, perspectives, and prospects. Science 349(6245):255–260
27. Jung TJ (2019) How transparency through blockchain helps the cybersecurity community. IBM Supply Chain and Blockchain Blog-Blockchain Explained
28. Ko HS, Nof SY (2012) Design and application of task administration protocols for collaborative production and service systems. Int J Prod Econ 135(1):177–189
29. Liu W, Gu W, Sheng W, Meng X, Wu Z, Chen W (2014) Decentralized multi-agent system-based cooperative frequency control for autonomous microgrids with communication constraints. IEEE Trans Sustain Energy 5(2):446–456
30. MIT Technology Review Editors (2018) Explainer: what is a blockchain? MIT Technology Review. https://www.technologyreview.com/2018/04/23/143477
31. Moghaddam M, Nof SY (2022) Information flow optimization in augmented reality systems for production & manufacturing. 26th international conference on production research
32. Nair AS, Nof SY, Bechar A (2021) Emerging directions of precision agriculture and agricultural robotics, pp 177–210
33. Nguyen WPV, Dusadeerungsikul PO, Nof SY (2022) Plant stress propagation detection and monitoring with disruption propagation network modelling and Bayesian network inference. Int J Prod Res 60(2):723–741
34. Nikolakis N, Maratos V, Makris S (2019) A cyber physical system (CPS) approach for safe human-robot collaboration in a shared workplace. Robot Comput Integr Manuf 56:233–243
35. Nof SY (2023) Automation: what it means to us around the world, definitions, its impact, and outlook, pp 3–38
36. Nof SY, Ceroni J, Jeong W, Moghaddam M (2015) Revolutionizing collaboration through e-work, e-Business, and e-Service (1st ed.)
37. Nof SY, Silva JR (2018) Perspectives on manufacturing automation under the digital and cyber convergence. Polytechnica 1(1–2):36–47
38. Oussous A, Benjelloun F-Z, Ait Lahcen A, Belfkih S (2018) Big data technologies: a survey. J King Saud Univ Comput Inf Sci 30(4):431–448
39. Reyes Levalle R (2018) Resilience by teaming in supply chains and networks (Springer ACES Series, Vol 5). Springer International Publishing
40. Reyes Levalle R, Nof SY (2017) Resilience in supply networks: definition, dimensions, and levels. Annu Rev Control 43:224–236
41. Reyes Levalle R, Scavarda M, Nof SY (2013) Collaborative production line control: minimisation of throughput variability and WIP. Int J Prod Res 51(23–24):7289–7307
42. Rosegrant MW, Cline SA (2003) Global food security: challenges and policies. Science 302(5652):1917–1919
43. Salah K, Rehman MHU, Nizamuddin N, Al-Fuqaha A (2019) Blockchain for AI: review and open research challenges. IEEE Access 7:10127–10149
44. Sharp M, Ak R, Hedberg T (2018) A survey of the advancing use and development of machine learning in smart manufacturing. J Manuf Syst 48:170–179
45. Sreeram M, Nof SY (2021) Human-in-the-loop: role in cyber physical agricultural systems. Int J Comput Commun Control 16(2)
46. Ullah Z, Al-Turjman F, Mostarda L, Gagliardi R (2020) Applications of artificial intelligence and machine learning in smart cities. Comput Commun 154:313–323
47. Villanueva A, Zhu Z, Liu Z, Wang F, Chidambaram S, Ramani K (2022) ColabAR: a toolkit for remote collaboration in tangible augmented reality laboratories. Proc ACM Human-Comput Interact 6(CSCW1):1–22

48. Wang N, Zhang N, Wang M (2006) Wireless sensors in agriculture and food industry—recent development and future perspective. Comput Electron Agric 50(1):1–14
49. Yang T, Huang Z, Pen H, Zhang Y (2017) Optimal planning of communication system of CPS for distribution network. J Sensors 2017:1–10

Glossary

A/DN Associate/Dissociate with a Network
AGV Automated Guided Vehicle
AR Augmented Reality
ARS Agricultural Robotic System
$ARS_{Group}(t)$ Group of ARS agents at time t
$ARS_{Single}(t)$ ARS single agent at time t
AS Adaptive Search algorithm
BMP Best Match Protocol
C&E Conflicts and Errors
C_I Inspection cost
C_{max} Makespan
C_O Overage cost
C_U Underage cost
CCC Collaboration, Coordination, and Control
CCLP Cyber-Collaborative Learning Protocol for Precision Agriculture
CCOSP Cyber-Collaborative Optimal Search Protocols for Precision Agriculture
CCP Cyber-Collaborative Protocol
CCP-CPS Cyber-Collaborative Protocol for Cyber-Physical System
CCP-ED Cyber-Collaborative Protocol for Early Detection of Stress in Crop Plants
CCP-SoS Cyber-Collaborative Protocol for System-of-Systems
CCT Collaborative Control Theory
CDF Cumulative Density Function
CI Collaborative Intelligence
CIM Computer Integrated Manufacturing
CPS Cyber-Physical System
CRDP Collaboration in Response to Disruption Propagation
CRM Collaboration Requirement Matrix
CRP Collaboration Requirement Planning
CRP-I Optimizer

P. O. Dusadeerungsikul and S. Y. Nof, *Cyber-Collaborative Algorithms and Protocols*, Automation, Collaboration, & E-Services 15,
https://doi.org/10.1007/978-3-031-56196-2

CRP-II Harmonizer
CRS Collaborative Robots Scheduling algorithm
d_i Number of infected crop plants found at location i
D-AS Dynamic Adaptive Search algorithm
DCS Demand and Capacity Sharing
DLOC Dynamic Lines of Collaboration
DSS Decision Support System
E Set of potential propagation edge
ECDP Error & Conflict Detection & Prevention
EPCR Error Prevention and Conflict Resolution
ERP Enterprise Resource Planning
EWP e-Work Parallelism
FCFS First Come, First Serve
FDL Facility Description Language
FN False Negative
FoF Factories of the Future
FP False Positive
FTT Fault Tolerance by Teaming
FTTP Fault-Tolerant Time-Out Protocol
G Set of two-dimensional grid
GCEEC Gateway Clustering Energy-Efficient Centroid
HITL Human-in-the-loop
HUB-CIs Hubs of Collaborative Intelligence
IoT/IoS Internet of Things/Internet of Services
K Maximum load for robot or team of robots
L Length dimension
LAN Local Area Network
m Task number m
M Set of crop plant
MDR-CPS Monitoring, Detecting, and Responding – Cyber-Physical System framework
ML&AI Machine Learning and Artificial Intelligence
MP Monitoring Profit
MP_S Monitoring Profit from discovering crop plant stress
MP_H Monitoring Profit from discovering healthy crop plant
n Resource number n
N_m Number of tasks on type
m
N_R Number of robot or team of robots
$N_r(t)$ Set of constraints, r, at time t
N_T Total number of tasks
$NICF$ Number of Infected Crop Plants Found
$O(m)$ Inspection status of plant m
O_b Maximum number of inspections allowed
$O_{BayesNet}$ Bayesian Network-Driven Scanning Algorithm

$O_{DualDegree}$ Value-based Dual-degree Scanning Algorithm
$O_{InDegree}$ Value-based In-degree Scanning Algorithm
O_{Random} Random Sampling Scanning Algorithm
O_{Value} Value-based Scanning Algorithm
ORE Overall Robotic Effectiveness
P Performance
p Probability that the crop plant develops stress
p_0 Stress origin probability
p_m Processing time of task m
PA Precision Agriculture
PCol Precision Collaboration
p_i Inspection time at location i
q Probability that stress propagates in a specific direction
q_i Searching time at location i
R Rejection region R
Layer Resource layer
R_0 Critical ratio
R_i Robot i
$R_i + R_j$ Team of robot i and j
RSR Redundant Scanning Ratio
S Success
$S(m)$ Crop plant's final stress status
$S_0(m)$ Origin stress status
$S_{ARS_{Group}}(t)$ State of the group of ARS agents at time t
$S_{ARS_{Single}}(t)$ State of the ARS agent at time t
S-AS Static Adaptive Search Algorithm
SD Standard Deviations
s_i Number of inspected crop plants at location i
SP, SP' Number of sampled locations in the greenhouse
SPM Stress Propagation Model
ST Crop plant has stress
ST' Crop plant does not have stress
t Time t of the system
T Total available time T
Layer Task layer
t_{ij} Travel time from i to j
TAP Task Administration Protocol
TCP/IP Transmission Control Protocol/Internet Protocol
TN True Negative
TNR True Negative Ratio
TOP Time-Out Protocol
TP True Positive
TPR True Positive Ratio
U Utilization
UR Unplanned Requests

USR Undetected Stresses Ratio
VR Virtual Reality
W Width dimension
w_m Priority of task m
WAN Wide-Area Network
WLAN Wireless Local Area Network
WSN Wireless Sensor Network
WSPT Weighted Shortest Processing Time First algorithm
$Y(x)$ Result from assumed parameters
$Y(x^*)$ Optimal result
α Type 1 error
β Type 2 error
γ Set of transition stages
γ_s Transition stage
δ Set of dependency
δ_{ijk} Task dependency
π Set of participants or agents
π_i Autonomous initiator agents
π_r Autonomous responder agents
σ Set of coordination/workflow protocols
σ_p Coordination protocol
τ Set of problems or tasks
τ_a Task
φ Set of activities of each agent
φ_k Availability indicator of the agent at a particular time